Springer Tracts in Modern Physics 103

Springer Tracts in Modern Physics

79 **Elementary Particle Physics** With contributions by E. Paul, H. Rollnick, P. Stichel

80* **Neutron Physics** With contributions by L. Koester, A. Steyerl

81 **Point Defects in Metals I:** Introductions to the Theory (2nd Printing)
By G. Leibfried, N. Breuer

82 **Electronic Structure of Noble Metals, and Polariton-Mediated Light Scattering**
With contributions by B. Bendow, B. Lengeler

83 **Electroproduction at Low Energy and Hadron Form Factors**
By E. Amaldi, S. P. Fubini, G. Furlan

84 **Collective Ion Acceleration** With contributions by C. L. Olson, U. Schumacher

85 **Solid Surface Physics** With contributions by J. Hölzl, F. K. Schulte, H. Wagner

86 **Electron-Positron Interactions** By B. H. Wiik, G. Wolf

87 **Point Defects in Metals II:** Dynamical Properties and Diffusion Controlled Reactions
With contributions by P. H. Dederichs, K. Schroeder, R. Zeller

88 **Excitation of Plasmons and Interband Transitions by Electrons** By H. Raether

89 Giant Resonance Phenomena in **Intermediate-Energy Nuclear Reactions**
By F. Cannata, H. Überall

90* **Jets of Hadrons** By W. Hofmann

91 **Structural Studies of Surfaces**
With contributions by K. Heinz, K. Müller, T. Engel, and K. H. Rieder

92 **Single-Particle Rotations in Molecular Crystals** By W. Press

93 **Coherent Inelastic Neutron Scattering in Lattice Dynamics** By B. Dorner

94 **Exciton Dynamics in Molecular Crystals and Aggregates** With contributions by
V. M. Kenkre and P. Reineker

95 **Projection Operator Techniques** in Nonequilibrium Statistical Mechanics
By H. Grabert

96 **Hyperfine Structure in 4d- and 5d-Shell Atoms** By S. Büttgenbach

97 **Elements of Flow and Diffusion Processes in Separation Nozzles** By W. Ehrfeld

98 **Narrow-Gap Semiconductors** With contributions by R. Dornhaus, G. Nimtz, and
B. Schlicht

99 **Dynamical Properties of IV–VI Compounds** With contributions by H. Bilz, A. Bussmann-
Holder, W. Jantsch, and P. Vogl

100* **Quarks and Nuclear Forces** Edited by D. C. Fries and B. Zeitnitz

101 **Neutron Scattering and Muon Spin Rotation** With contributions by R. E. Lechner,
D. Richter, and C. Riekel

102 **Theory of Jets in Electron-Positron Annihilation** By G. Kramer

103 **Rare Gas Solids** With contributions by H. Coufal, E. Lüscher, H. Micklitz, and
R. E. Norberg

104 **Surface Enhanced Raman Vibrational Studies at Solid/Gas Interfaces** By I. Pockrand

105 **Two-Photon Physics at e^+e^- Storage Rings** By H. Kolanoski

---

* denotes a volume which contains a Classified Index starting from Volume 36.

Rare Gas Solids

Contributions by
H. Coufal   E. Lüscher   H. Micklitz
R. E. Norberg

With 40 Figures

Springer-Verlag Berlin Heidelberg GmbH 1984

Hans Coufal
Professor Dr. Edgar Lüscher

Physikdepartment E13, Technische Universität München
D-8046 Garching, Fed. Rep. of Germany

Professor Dr. Hans Micklitz

Institut für Experimentalphysik IV, Ruhr Universität Bochum
D-4630 Bochum, Fed. Rep. of Germany

Professor Dr. Richard E. Norberg

Washington University, Department of Physics
St. Louis, MO 63130, USA

Manuscripts for publication should be addressed to:

Gerhard Höhler

Institut für Theoretische Kernphysik der Universität Karlsruhe
Postfach 6380, D-7500 Karlsruhe 1, Fed. Rep. of Germany

*Proofs and all correspondence concerning papers in the process of publication
should be addressed to:*

Ernst A. Niekisch

Haubourdinstrasse 6, D-5170 Jülich 1, Fed. Rep. of Germany

ISBN 978-3-662-15724-4   ISBN 978-3-540-38835-7 (eBook)
DOI 10.1007/978-3-540-38835-7

Library of Congress Cataloging in Publication Data. Main entry under title: Rare gas solids. (Springer tracts in modern physics; 103) Includes bibliographies and indexes. 1. Solid rare gases – Addresses, essays, lectures. I. Coufal, H. II. Series. QC1.S797 vol. 103   539 s [546'.75]   84-5610 [QC176.8.R35]

Preface

Spectroscopic studies or rare-gas-matrix-isolated (RGMI) species have been performed for almost thirty years. Electronic spectroscopy was the first technique used for such studies. Electron spin resonance (ESR) spectroscopy entered this field about ten years later and it took almost another ten years before the first nuclear gamma resonance (NGR), or Mössbauer effect, experiment on a RGMI species was reported. Originally the possibility of studying otherwise unstable molecules or radicals was the main achievement of the RGMI technique. During the years, matrix-isolation work shifted more and more towards stable molecules and their aggregates, since it was realized that important information on the structure and interactions in these species can be obtained from such experiments. With the introduction of new techniques such as ESR and NGR spectroscopy it was now possible to study the hyperfine interaction in RGMI species. This opened a new field: hyperfine studies of "almost free" atoms or ions. With the help of ESR spectroscopy on RGMI atoms or ions, one could study the hyperfine interaction of the electronic ground state of the prototype of an impurity in a solid: a single paramagnetic atom or ion in a matrix with a very weak interaction between the atom or ion and the surrounding matrix. The spectra should be easily understood starting from basic quantum mechanics and solid-state principles. The Mössbauer effect, on the other hand, is intrinsically a solid-state effect since it is based on the interaction of the resonating nucleus with the lattice. NGR studies of RGMI atoms or ions, i.e., on very weakly bound atoms or ions, offer the unique opportunity to measure the electron density at the resonating nucleus of "almost free" atoms or ions via the Mössbauer isomer shift. Such information is not obtainable by any other technique.

Nuclear magnetic resonance (NMR) measurements in rare gas solids and liquids have been made for nearly twenty five years. The initial emphasis of such studies was on determinations of coefficients of atomic self-diffusion and of the large temperature-dependent resonance shifts. Since three of the four commonly employed magnetic nuclides have nuclear spin greater than $\frac{1}{2}$, quadrupolar effects were found to play a significant role in the experimental results. A concurrent remarkable development of the theory of dynamic quadrupolar phenomena has produced an extremely detailed understanding of some of the relaxation and diffusion results. The principal significant advance in relevant experimental methods over this period has been the introduction of adiabatic demagnetization in the rotating reference frame (ADRF) which has extended the NMR determinations of atomic diffusion to include ultra-slow motions (for example, with characteristic times of thousands of seconds). The availability of computer-controlled NMR spectrometers based on stable and intense fields from superconducting solenoids has made possible extensive data accumulations on weakly magnetic nuclear spin systems. In recent years NMR study of molecular hydrogen in rare gas solids and liquids has been very fruitful and has produced unique determinations of the temperature variation of the correlation function for molecular relaxation of dilute ortho-H_2. This work now has proved to be relevant for the understanding of the presence and dynamics of molecular hydrogen in other materials, such as amorphous silicon.

In this volume we review the NMR studies in condensed rare gases and ESR and NGR studies of RGMI atoms and ions which have been performed so far. The information which could be obtained from these experiments is discussed in detail. We hope that this review will stimulate new activities in this field.

It is a pleasure for us to acknowledge collaboration, conversation and correspondence with many colleagues working in these fields. One of us (H. Micklitz) wants to express his special thanks to P.H. Barrett, University of California, Santa Barbara, for introducing him to the NGR-RGMI technique many years ago. Part of the manuscript was prepared while one of us (H. Coufal) was with the IBM Research Laboratory at San Jose. He is particularly grateful for helpful discussions with J. Pacansky and to T.R. Koehler for so carefully reading the manuscript. R.E.N. wishes particularly to acknowledge his debt to Peter A. Fedders for many illuminating conversations and for his splendid contributions to the description of dynamic quadrupole effects.

San Jose, München, Bochum   *H. Coufal*
St. Louis, March 1984   *E. Lüscher*
                                                           H. Micklitz
                                                          R.E. Norberg

Contents

Electron Spin and Nuclear Gamma Resonance Studies of Rare Gas Matrix-Isolated Atoms and Ions

By H. Coufal, E. Lüscher, and H. Micklitz (With 16 Figures)

1. Introduction 1

2. Experimental Technique 5
   2.1 Sample Preparation 5
       2.1.1 Codeposition 6
       2.1.2 In-situ Generation 7
       2.1.3 Ion Implantation 9
   2.2 Specific Problems in Nuclear Gamma Resonance Experiments 10
       2.2.1 Experimental Setup 10
       2.2.2 Sample Preparation 11
   2.3 Specific Problems in Electron Spin Resonance Experiments 13
       2.3.1 Sample Preparation 13
       2.3.2 Experimental Setup 15

3. Nuclear Gamma Resonance Experiments 16
   3.1 Experiments on RGMI Atoms 16
       3.1.1 ^{119}Sn 16
       3.1.2 ^{57}Fe 17
       3.1.3 ^{151}Eu 19
       3.1.4 ^{125}Te 21
   3.2 NGR Experiments on RGMI Ions 21
       3.2.1 ^{57}Fe^{+}($3d^{6}4s^{6}$D) 21
       3.2.2 ^{57}Fe^{+*}($3d^{7}$) 23
       3.2.3 ^{119}Sn^{+}($4d^{10}5s^{2}5p$) 24
   3.3 Discussion of Experimental Results 25
       3.3.1 Isomer Shift 25
       3.3.2 Hyperfine Coupling Constant 27
       3.3.3 Lamb-Mössbauer Factor 29

      3.3.4 Spin-lattice Relaxation 31

      3.3.5 Crystal Field Parameters 31

   3.4 Summary 33

4. ESR Experiments 33

   4.1 Paramagnetic Impurities with S Ground State Symmetry 35

      4.1.1 Atoms with $(ns)^1$ Electron Configuration 35

      4.1.2 Atoms with $(np)^3$ Electron Configuration 42

      4.1.3 Atoms and Ions with $(3d)^5(4s)^n$ Electron Configuration 44

   4.2 Paramagnetic Impurities with P Ground State Symmetry 45

   4.3 Double Resonance Experiments 48

   4.4 Models for the Interpretation of the ESR Spectra of RGMI Atoms and Ions 51

      4.4.1 Models for Impurities with S Ground State Symmetry 51

      4.4.2 Models for Impurities with P Ground State Symmetry 54

   4.5 Summary 54

References 55

Combined Subject Index 97

Nuclear Magnetic Resonance in Condensed Rare Gases
By R.E. Norberg (With 24 Figures)

1. Introduction 59

2. Neon, Krypton, and Xenon 59

   2.1 Spin Relaxation and Diffusion 59

      2.1.1 Rare Gas Solids 60

      2.1.2 Rare Gas Liquids 74

   2.2 Chemical Shifts 80

3. Dilute H_2 in Neon, Argon, and Krypton 84

   3.1 H_2 in Rare Gas Liquids 84

      3.1.1 Diffusion 84

      3.1.2 Relaxation Times 86

   3.2 H_2 in Rare Gas Solids 87

4. Concluding Remarks 92

References 93

Combined Subject Index 97

List of Abbreviations

ADRF   Adiabatic Demagnetization in the Rotating Reference Frame
CF   Crystal Field
EFG   Electric Field Gradient
ESR   Electron Spin Resonance
hf   hyperfine
ID   Inner Diameter
IS   Isomer Shift
MIS   Matrix Isolation Spectroscopy
NGR   Nuclear Gamma Resonance
NMR   Nuclear Magnetic Resonance
RG   Rare Gas
RGMI   Rare Gas Matrix Isolated
SCF   Self Consistent Field
STP   Standard Pressure
SVP   Saturated Vapor Pressure

Electron Spin and Nuclear Gamma Resonance Studies
of Rare Gas Matrix-Isolated Atoms and Ions

By H. Coufal, E. Lüscher, and H. Micklitz

1. Introduction

Matrix-Isolation-Spectroscopy (MIS) measurements on impurities isolated in a solid
inert matrix is frequently used to analyze reagents that otherwise would be unstable.
Due to the development of small self-contained cryogenic refrigerators MIS has become, within the last few years, a well established spectroscopic technique, easily
accessible not only to physicists but also to chemists. In effect, the rigid, chemi-
cally inert, rare gas matrix became just another solvent for chemists, albeit a very
special one, that allowed new chemical reactions and the synthesis and the analysis
of unknown compounds and radicals (see for example, Moskovits and Ozin /1976/). Nu-
merous low temperature Electron Spin Resonance (ESR) experiments on radicals isolated
and stabilized in rare gas matrices are a typical example. For most of these experiments on molecules the interaction of the impurities with the rare gas matrix can be
completely neglected. That is why experiments on molecular impurities, for which one
can neglect the rare gas aspects of the matrix, are not discussed within this volume
on rare gas (RG) solids. Such experiments are discussed in the standard literature on
ESR and Mössbauer Spectroscopy on molecules in solid solvents.

   ESR experiments on rare gas matrix isolated atoms and ions have been conducted for
various reasons. At the beginning /Jen et al., 1958/ results on isolated atoms were a
byproduct of research on radicals; this is especially true for hydrogen atoms. Hydrogen is a constituent of most matrix isolated molecules and by cracking these molecules
isolated hydrogen atoms are created. Another important motivation was the study of the
products of electrical discharges in gases /Wall et al., 1959a,b/ or γ irradiation of
solids /Bouldin and Gordy, 1964/. Observing optical pumping phenomena at very high buf-
fer gas pressures in a solid matrix was another aspect /Kupfermann and Pipkin, 1958/.
The main goal, however, for ESR experiments on matrix isolated atoms and ions was the
study of the electronic ground state of the prototype of an impurity in a solid: a
single paramagnetic atom or ion in an inert rare gas matrix. Due to the simple structure of these impurities and their weak interaction with the respective matrices their
spectra should be easily understood starting from first quantum mechanical and solid

1

state principles. The situation is quite different for Nuclear Gamma Resonance (NGR). The Mössbauer effect is intrinsically a solid state effect. The matrix and its interaction with the nuclei that are studied are essential in order to give a non-zero recoilless fraction. This interaction, however, is weak in RG solids and its influence on the hyperfine (hf) parameters of the impurity is expected to be small or even negligible. In this way the hf interaction of "almost free" atoms or ions can be studied in RGMI-NGR experiments. The idea of Mössbauer experiments on Rare Gas Matrix Isolated (RGMI) atoms and ions dates back to the year 1962 when Jaccarino and Wertheim /1962/ suggested such experiments for the purpose of the Mössbauer Isomer Shift (IS) calibration. However, it was not before 1971 when the first successful Mössbauer experiments on RGMI ^{57}Fe atoms were reported /McNab et al., 1971a,b/. In the meantime these RGMI Mössbauer experiments have been extended to other isotopes and charged species. While the main purpose of these experiments was the determination of IS calibration points other important information has been obtained. These are: the magnetic hyperfine field (A factor) at the Mössbauer nucleus for well defined atomic configurations, nuclear quadrupole moments from the quadrupolar interaction in metallic dimers (Fe_2 and Sn_2), information about lattice dynamics in doped rare gas solids (from the temperature dependence of the Mössbauer-Lamb-Factor or from the temperature dependence of the spin-relaxation time of magnetic atoms) and finally the order of magnitude and sign of crystal field parameters for atomic or ionic ground states. A review on the Mössbauer effect data obtained from experiments on matrix isolated species (atoms, ions, molecules and clusters) has been given recently /Micklitz, 1981/.

Compared to most optical techniques NGR- and ESR-spectroscopy are two more recent and sensitive spectroscopic techniques: for a typical Mössbauer experiment 10^{16} nuclei are necessary, whereas for ESR 10^{13} paramagnetic impurities are sufficient. Mössbauer spectroscopy is a nuclear resonance technique, whereas ESR observes resonance transitions between electronic levels. Both techniques however are related by the hyperfine interaction between the nucleus of an atom and its electrons. Thus the electronic ground state of a matrix isolated atom can be directly analyzed by ESR but also indirectly by NGR. ESR is able to determine the resulting electron spin density at the nucleus of an atom, and NGR measures in addition the electron charge density there. This demonstrates that despite many differences between NGR and ESR, especially experimental differences, both effects can be described by one common Hamiltonian operator as far as the properties of RGMI atoms and ions are concerned.

The Hamiltonian $\hat{H}$ of an RGMI impurity may be written in hierarchic order in the following way:

$$\hat{H} = \hat{H}_{Coul} + \hat{H}_{LS} + \hat{H}_{CF} + \hat{H}_{ext}^m + \hat{H}_{hf} \quad . \tag{1.1}$$

We want to discuss now in detail each of these five contributions a) - e) to $\hat{H}$ in (1.1):

a) $\hat{H}_{Coul}$ is the Coulomb interaction between all electrons and the point nucleus of the impurity,

b) $\hat{H}_{LS} = \lambda \mathbf{L} \cdot \mathbf{S}$ is the spin-orbit interaction of the valence electrons of the impurity,

c) $\hat{H}_{CF}$ is the crystal field interaction of the matrix with the impurity.
For the case that J is a good quantum number (see below) and for cubic symmetry at the impurity lattice site, we have

$$\hat{H}_{CF}^{cubic} = \Delta[35\hat{J}_z^4 - 30J(J+1)\hat{J}_z^2 + 25\hat{J}_z^2 - 6J(J+1) + 3\hat{J}^2(J+1)^2 + \tfrac{5}{2}(\hat{J}_+^4 + \hat{J}_-^4)] \qquad (1.2)$$

where Δ is a measure of the cubic field strength. If the impurity lattice site shows an axial distortion, we have

$$\hat{H}_{CF} = \hat{H}_{CF}^{cubic} + \hat{H}_{CF}^{axial} \qquad (1.3)$$

with

$$\hat{H}_{CF}^{axial} = \delta[3\hat{J}_z^2 - J(J+1)] \qquad (1.4)$$

where δ is a measure of the axial field strength.

Usually the contribution due to the crystal field is small compared to the spin-orbit interaction, $\hat{H}_{CF} \ll \hat{H}_{LS}$. Therefore J is a good quantum number and the effect of the crystal field can be described by lifting the (2J+1)-fold degeneracy of the atomic or ionic multiplet levels. These crystal field states can be treated in the spin Hamiltonian formalism by introducing an effective spin S_{eff} and an effective g tensor g_{eff}. In ESR- and NGR-experiments with RGMI impurities one studies the Zeeman and hyperfine interaction of these crystal field states (characterized by S_{eff} and g_{eff}). Only in the special case of non-degenerate, atomic S states, of the impurity (for example alkali atoms) one studies the Zeeman and hyperfine interaction of the free atomic state described by J and g.

d) The Zeeman interaction $\hat{H}_{ext}^m$ of the impurity in an external magnetic field B_{ext} can be separated in an electronic and a nuclear part:

$$\hat{H}_{ext}^m = - g_{eff}\beta_e \hat{S}_{eff} \cdot \hat{B}_{ext} - g_N \beta_N \hat{I} \cdot \hat{B}_{ext} \qquad (1.5)$$

with β_e and β_N the Bohr and the nuclear magneton respectively, g_N the nuclear g factor and I the nuclear spin. $\hat{H}_{ext}^m$, usually not present in NGR experiments since most

3

of these experiments are done without an external magnetic field, is essential for all ESR experiments.

e) $\hat{H}_{hf}$ is the hyperfine interaction of the impurity. A multipole expansion of $\hat{H}_{hf}$ and separation of magnetic $\hat{H}_{hf}^m$ and electric $\hat{H}_{hf}^e$ contributions gives:

$$\hat{H}_{hf} = \hat{H}_{hf}^m + \hat{H}_{hf}^e \qquad (1.6)$$

with

$$\hat{H}_{hf}^m = \hat{H}_{hf}^{M_1} + \dots \quad \hat{=} \text{ magnetic dipole interaction} + \dots \qquad (1.7)$$

$$\hat{H}_{hf}^e = \hat{H}_{hf}^{E_0} + \hat{H}_{hf}^{E_2} + \dots \quad \hat{=} \text{ correction to } \hat{H}_{coul} \text{ due to an extended nucleus} \qquad (1.8)$$
$$\text{+ electric quadrupole interaction} + \dots$$

These terms are usually written in the form

$$\hat{H}_{hf}^{M_1} = A_{eff} \cdot \hat{I} \cdot \hat{S}_{eff} = A \cdot \frac{g_{eff}}{g} \cdot \hat{I} \cdot \hat{S}_{eff} \qquad . \qquad (1.9)$$

A is the so-called hf-coupling tensor; in the case of axial symmetry (for example, in the presence of an applied external field B_{ext}) $\hat{H}_{hf}^{M_1}$ can be written as

$$\hat{H}_{hf}^{M_1} = g_N \cdot \beta_N \cdot \hat{I} \cdot \hat{B}_{eff} \qquad (1.10)$$

with

$$B_{eff} = A(g_{eff}/g \ g_N \beta_N) S_{eff} \qquad . \qquad (1.11)$$

$\hat{H}_{hf}^{M_1}$ is usually the only term in the hyperfine interaction which is measured both in ESR and NGR experiments on RGMI atoms or ions.

$$\hat{H}_{hf}^{E_0} = \frac{2\pi}{3} e^2 Z\rho(0)<r^2> \qquad . \qquad (1.12.)$$

$$\hat{H}_{hf}^{E_2} = \frac{e^2 Q}{4I(2I-1)} q_{zz}[3\hat{I}_z^2 - I(I+1) + \eta(\hat{I}_x^2 - \hat{I}_y^2)] \qquad , \qquad (1.13)$$

with $\eta = (q_{xx} - q_{yy})/q_{zz}$; $\rho(0)$ is the electron density at the nucleus, Z the nuclear charge, Q the nuclear quadrupole moment, $<r^2>$ the mean square radius of the nuclear charge distribution and q_{xx}, q_{yy} and q_{zz} are the components of the electric field gradient tensor at the nucleus.

$\hat{H}_{hf}^{E_0}$ gives rise to the so-called Mössbauer isomer shift IS $= (2\pi/3) e^2 Z\Delta\rho(0)\Delta<r^2>$ in NGR experiments due to the change $\Delta<r^2>$ in $<r^2>$ between the nuclear ground state

4

and the nuclear excited state; $\Delta\rho(0)$ is the difference in $\rho(0)$ between the Mössbauer source and absorber. This term in $\hat{H}_{hf}$ is not observable in ESR experiments, where there is no change in $<r^2>$.

$\hat{H}_{hf}^{E2}$ results in the quadrupole splitting ΔE_Q; it is only important if the RGMI atom or ion is on a non-cubic lattice site in the rare gas matrix, otherwise $q_{xx} = q_{yy} = q_{zz} = 0$, i.e., $\Delta E_Q = 0$.

Thus, in a typical ESR experiment on a RGMI impurity B_{ext} is used to measure $\hat{H}_{ext}^m$ and $\hat{H}_{hf}^{M1}$ with a relative high resolution and to obtain information about shifts in g and A due to the rare gas matrix. NGR experiments measure $\hat{H}_{hf}^{E0}$ and $\hat{H}_{hf}^{M1}$ with a relatively low resolution and give information about $\rho(0)$ and A. Since $\rho(0)$ is not accessible by ESR it is the most important quantity obtained from NGR-RGMI experiments.

2. Experimental Technique

Although standard ESR spectrometers could have been used for ESR experiments with RGMI atoms or ions, up to now most experiments have used highly sophisticated, specially disigned equipment. This situation arose mainly for two reasons: first, the problems of handling and keeping a sample over long periods at low temperatures; these are overcome now by the availability of small self-contained cryogenic refrigerators. Second, the low impurity concentrations needed for well isolated single impurities require very sensitive spectrometers. MIS stimulated the technical development of the very sensitive high resolution ESR spectrometers that became available commercially during the last few years. For the NGR experiments reported here standard NGR equipment has been used. If a suitable spectrometer and a low temperature refrigerator is within reach of a scientist today, the only problem left to tackle is the proper sample preparation.

2.1 Sample Preparation

Due to the properties of solid rare gases, MIS is dominated by the preparation of suitable samples at low temperatures. Several techniques are used to produce doped rare gas samples.

A doped matrix may be prepared from a corresponding gas mixture by depositing this gas mixture on a cold substrate or by growing a crystal from this gas mixture. Another approach would be to produce the sample by generating within a rare gas crystal by a chemical or physical reaction the species that shall be examined by MIS. Both techniques, the codeposition of matrix and impurity and the in situ generation of the impurity within the matrix, have been widely employed to produce isolated, single atoms or ions in a rare gas matrix. These samples are then used as an absorber with the usual ESR or NGR spectroscopy apparatus; in NGR the sample may be used as a source as well.

2.1.1 Codeposition

The standard procedure which is used for the sample preparation in RGMI-NGR experiments as well as for ESR is the simultaneous condensation of rare gas atoms and the impurity under study onto a cold (usually 6-10 K) substrate. In principle, this method allows any combination and concentration of dopant and rare gas. Therefore this technique dominates the MIS (e.g. /Meyer, 1971/). But it suffers from several disadvantages as far as studies on isolated atoms or ions are concerned:

1) Many atoms cluster easily in the gas phase, or due to diffusion, on the surface of the slowly growing sample.
2) Insufficient vacuum conditions cause undesired additional impurities. To overcome their influence often the largest possible doping concentration is selected thus introducing interactions between the atoms under study.

For a well defined sample therefore the conditions during and after the deposition play an important role. The matrix gas deposition rate is the most important factor in the sample preparation. Low deposition rates form a well annealed sample whereas high deposition rates cut down the clustering due to surface migration. Deposition rates in the order of 1 to 10 atomic layers per second seem to be a good compromise between these intrinsically opposite goals. Depending on the desired dopant concentration and the maximum allowable ratio of additional impurities this rare gas flow rate determines also dopant source and the necessary vacuum conditions.

High vacuum with a residual gas pressure better than 10^{-6} torr is in any case necessary, ultra high vacuum < 10^{-9} torr would be favorable for most experiments, for some this vacuum is even indispensable.

For the purpose of evaporating the dopant Knudsen cells have been widely used. The advantage to be gained by using these cells is that the atoms are in thermal equilibrium and it is possible to calculate the atomic flux knowing the temperature of the cell and its channel dimensions. Higher atomic flux can be achieved by using crucibles or hf discharges as sources though loosing the advantages of the Knudsen cell.

The fraction of monomers in the vapor phase of some metals is negligibly small at temperatures corresponding to a metal vapor pressure < 10^{-3} torr. The beam of such metals emerging from a Knudsen cell or an open ended crucible therefore contains essentially dimers, but not monomers. In order to obtain an atomic beam of such metals, one has to use either a double Knudsen cell arrangement or a Knudsen cell in connection with a discharge region. The first cell at a relatively low temperature produces a dimeric beam, the second cell at a considerably higher temperature or the discharge breaks the dimer bond.

Rare gas and dopant are deposited by spray nozzles, concentric tubes, various apertures and other devices on a cold surface. The substrate is in thermal contact with a helium bath or the cold end of a closed cycle refrigerator. Substrate materials should have a high thermal conductivity and should cause a resonance signal, i.e., sapphire is used for ESR studies and beryllium or aluminum foils are used for NGR studies.

2.1.2 In-situ Generation

The codeposition techniques dominate the field. But problems arise if the species to be studied is unstable in the vapor phase or during the deposition, as for example with ions. Special procedures have been developed for the sample preparation in these cases: procedures that are often favorable even for impurities that are accessible to the standard techniques.

Photochemical generation. One method which has been widely used in this type of MIS experiment is the matrix isolation of molecules containing the ion or atom to be studied followed by the photodissociation of the RGMI molecules. Frequently radiation from UV light sources (i.e. /Montano et al., 1976/) but also γ radiation from sources like Co^{60} (i.e. /Boulding and Gordy, 1964/) have been used to initiate the dissociation. As far as studies on isolated impurity atoms or ions are concerned this technique suffers, however, from a severe disadvantage: the recombination, that would take place immediately if the reaction products stay close to each other, has to be suppressed.

Hydrogen atoms are highly mobile in rare gas crystals above 20-40 K (see Sect. 4.1.1). Therefore, experiments on isolated hydrogen atoms /Bouldin and Gordy, 1964/ or on atoms or ions that are components of an hydride prior to dissociation are one way to avoid recombination problems. An example of the photolysis of a hydride is the NGR experiment on ^{125}Te by Montano et al. /1976/. They isolated H_2Te, which is highly unstable under normal conditions, in rare gas matrices; subsequent photodissociation of the RGMI H_2Te molecules produced atomic Te and hydrogen. The latter is highly mobile in RG matrices even at 4.2 K and leaves the RGMI Te atom with a slightly distorted rare gas surrounding.

Kasai /1968/ demonstrated another method to avoid the recombination problem. For experiments on RGMI metal ions he put HI molecules in the matrix in addition to the metal atoms M. UV irradiation of such a matrix produces now stable M^+ ions in a two step process:

1) Photodissociation of HI by the excitation of $^3\Pi$ and $^2\Pi$ states into H and I /Mulliken, 1937/.

2) Photoexcitation of M accompanied by a charge transfer process between the excited M* and the I atom

$$M* + I \rightarrow M^+ + I^- \quad .$$

This charge-transfer process seems to be an "energy resonance" process between the two electronic levels involved /Micklitz and Barrett, 1972c; Gerth et al., 1972/. According to Kasai /1968/ the return of the electron from the anion I^- to the cation M^+ is hampered by the local potential trap imposed by the electron affinity of the electrophilic species. The resulting ions I^- and M^+ are separated by an average distance <d> which is determined by the atomic concentrations of M and HI; typical concentrations of 1% HI and 0.1 to 0.5% M give <d> ~ 2 nm. Thus the electric field of the I^- will produce an electric field gradient at the location of M^+.

Kasai /1968/ showed in his ESR experiments that such RGMI ions are stable at least for hours, NGR experiments on $^{57}Fe^+$ in xenon matrices of Micklitz and Litterst /1974/ and of Montano et al. /1978/ indicate that these ions are stable even for weeks.

Up to now the UV photolysis of matrix isolated molecules is the only way to produce RGMI stable ions. However, interactions of the ions under study with undissociated molecules or parts of these have to be accepted.

Radiochemical generation. A method suggested by Lambe /1966/ avoids the problems of the photochemical generation of the dopant by generating it by a radiochemical process via nuclear instead of chemical reactions.

A first step by Micklitz and Luchner /1969/ was to condense a rare gas with a radioactive isotope which decays into the desired impurity and thus gives statistically distributed, isolated single impurities in ionized states. The β^- decay of ^{85}Kr gives, for example, $^{85}Rb^+$ in a ^{85}Kr matrix. If the lifetime of the excited nuclear state of this ion is shorter than its recombination time with electrons then Mössbauer *source* experiments of the ion are possible.

The radioactive decay can be a β decay, an electron capture decay or a converted nuclear γ transition. The latter two decays are followed by an Auger cascade. Pollack /1962/, for example, analyzed the Auger cascade in ^{57}Fe following the electron capture decay of free ^{57}Co atoms. This analysis shows that 94% of the ^{57}Fe impurities will be in ionized states up to $^{57}Fe^{6+}$. These ions are not stable against recombination with free electrons; Fe^+ in a xenon matrix has a recombination time of ~ $3 \cdot 10^{-7}$ s at 4.2 K /Micklitz and Barrett, 1972b/ and optical experiments on the recombination of $^{85}Rb^+$ /Micklitz and Luchner, 1969/ give a lower limit for the recombination time of 10^{-8} s. This indicates that in Mössbauer source experiments with RGMI atoms, which have a lifetime of the excited nuclear state $< 10^{-8}$ s, charged states of the daughter ion should be observable. Since charged states will give additional Mössbauer isomer shift calibration points (see Sect.3.3.1) such experiments have been proposed already in 1962 by Jaccarino and Wertheim for the solution of the isomer shift calibration problem for ^{57}Fe. The production of RGMI ions by radioactive decay, however, has disadvantages too. The electronic configuration of these ions depends on the specific recombination mechanism of the highly charged ions created initially by the Auger cascade and the lifetime of these ions is too short for ESR or Mössbauer *absorber* experiments.

An additional step compared to the codeposition of the rare gas with a radioactive isotope is the neutron transmutation doping of a pure rare gas crystal via a (n,β^-) process.

By irradiation at 4.6 K of a pure argon crystal with thermal neutrons, a statistically distributed percentage of ^{40}Ar atoms (natural argon gas is 99.6% ^{40}Ar) is converted to ^{41}Ar by neutron capture. These ^{41}Ar atoms decay via a β decay with a lifetime of 1.8 hr into potassium ions that recombine immediately with free electrons. Thus an argon crystal can be doped via a $^{40}Ar (n,\beta^-)^{41}K$ process with isolated ^{41}K atoms that can be examined by ESR /Coufal et al., 1974b/. The advantage of this technique is that a well defined sample can be produced:

1) highest purity rare gas matrix with a single isotope as a dopant,
2) statistical distribution of single dopant atoms across the complete sample,
3) from the irradiation data the dopant concentration can be calculated precisely and
4) the chemical purity of the doped sample can easily be proven by γ activation analysis.

The disadvantage of this doping method is that it is practically restricted to argon crystals. But it's obvious benefits make the system $^{40}Ar:^{41}K$ a model system for the study of RGMI single metal atoms /Coufal et al., 1978/.

2.1.3 Ion Implantation

Mössbauer source experiments on RGMI species require the evaporation of radioactive materials which has the consequence of a contamination of evaporation system and cryostat. The ion implantation into pure RG matrices with the help of an ion accelerator which has been reported recently /Van Rossum et al., 1980/ seems to be a superior method for the sample preparation in Mössbauer source experiments: a ^{57}Co doped xenon matrix was obtained by ^{57}Co ion implantation, using an ion accelerator with an accelerating voltage of 85 kV; the ion dose was approximately 10^{14} at/cm^2, giving a Co concentration of $5\cdot10^{-5}$ in a layer of $\simeq$ 150 nm thickness.

The ion implantation method does not show the disadvantages of the codeposition method (residual gas impurities, clustering effects, surface migration). This technique may become important for RGMI studies of such species which can not be evaporated out of a furnace.

2.2 Specific Problems in Nuclear Gamma Resonance Experiments

2.2.1 Experimental Setup

Figure 2.1 shows the arrangement of the helium cryostat with the evaporation system and the Mössbauer spectrometer as it has been used for most of the RGMI experiments reported here. The alumina or quartz crucible is inserted into a tantalum furnace which can be resistance heated to ~1800 K. A tantalum shield reflects radiant energy back to the furnace and an outer water-cooled copper shield removes excess heat up to a rate of ~350 J/s. The material evaporated from the crucible will condense onto a low-iron beryllium disk (Mössbauer absorption due to the iron content in the beryllium disk ~0.08%) or an aluminum foil, which is held in an oxygen-free copper clamp thermally bonded to the bottom of the coolant bath (usually liquid helium). The RG enters with an adjustable flow rate and condenses onto the cold substrate together with the metal atoms.

The stainless steel cryostat sits on a rotatable "O"-ring seal so that outgassing of the furnace can be carried out before exposing the substrate to the matrix gas/

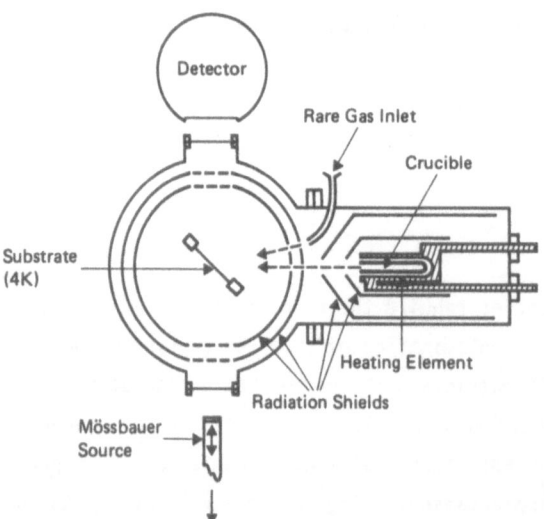

Fig.2.1. Experimental setup for Mössbauer experiments on RGMI impurities

metal atom mixture. Instead of rotating the cryostat a movable shutter between the furnace and the substrate can be built in for the purpose of outgassing the furnace. The gas handling lines are constructed of stainless steel tubing. The RG used in the experiments reported here have only a few ppm impurities.

For Mössbauer experiments which demand a cooled Mössbauer source (small Lamb-Mössbauer factor at room temperature) a modified version of the setup as shown in Fig. 2.1 has to be used /Montano et al., 1976/. The Mössbauer source and driving unit are built into the helium cryostat and the source is thermally connected by a thin copper wire to the helium bath.

A somewhat different setup has been used by Shamay and Pasternak /1976/ for Mössbauer experiments on RGMI alkali iodide molecules. This setup can be generally used for Mössbauer experiments with RGMI species which evaporate at relative low temperatures (< 1300 K). It should be mentioned, however, that due to the rotable feedthroughs used in this setup the residual vapor pressure in the helium cryostat was not better than $\sim 10^{-6}$ torr. This may cause impurity problems (see above) if such a setup is used for the matrix isolation of metal atoms.

2.2.2 Sample Preparation

After outgassing of the furnace, the cryostat is either rotated so that the three beam apertures are concentric (see Fig.2.1) or the shutter between the furnace and the substrate is opened. Initially, a layer of matrix gas (~ 50 $\mu g/cm^2$) is laid down on the substrate in order to eliminate metal atom substrate interaction. The furnace temperature is raised to the required value which in conjunction with the RG flow rate controls the RG to metal atomic ratio. The temperature is held to within ± 5 K and is read either with a thermocouple or an optical pyrometer or a calibrated solar cell via a window positioned at the rear of the furnace. For an absorber consisting of 1-2% metal atoms, the matrix gas enters the cryostat at a rate of ~ 1 cm^3 STP per minute; a sticking coefficient for the RG at 4.2 K of ~ 0.5 seems to be indicated. The matrix gas accretion on the substrate is continuously monitored by measuring the attenuation of the iron 6.4 keV X ray or the 14.4 keV γ ray from a ^{57}Co source. In this way the matrix thickness can be measured to within $\pm 10\%$. In the experiments reported here it was not possible, however to determine simultaneously the deposition rate of the metal atoms. This has been carried out in a separate experiment where a known weight of the metal is evaporated and the collection efficiency of the substrate (= amount of material on 1 cm^2 of substrate/total evaporated material) is measured by weighing the substrate before and after the deposition. The collection efficiency with the apparatus used for the experiments here was 1-2%. Thus weighting the crucible before and after a run together with the once determined collection efficiency gives the amount of metal in the absorber.

The method described above for the determination of sample thickness and metal atomic concentration has been used in all the RGMI Mössbauer experiments reported so far. However, the determination of deposition rates with the use of crystal quartz monitors (for example /Ryberg and Hunderi, 1977/) will certainly be the method of the future.

There is one important difference between Mössbauer and ESR spectroscopy with RGMI atoms: the Mössbauer spectra of RGMI atoms are much more "sensitive" to impurities and migration effects than the ESR spectra. The reason for this is the low resolution of Mössbauer spectroscopy in comparison to the high resolution of ESR spectroscopy. The ESR spectrum of an isolated atom is usually well separated from that of aggregates of atoms (dimer, trimer, etc.) or that of molecular species built by the atom under study and impurities. These aggregation and impurity effects are usually observed as just a decrease in the isolated atom absorption signal. In Mössbauer spectroscopy, however, these effects are a substantial disturbance since these aggregate and impurity spectra are difficult to separate from the isolated atom spectrum. It is for this reason that the first successful Mössbauer experiment on RGMI *atoms* was not reported before 1971, almost 20 years after the first optical and ESR-RGMI experiments.

For many years ESR spectroscopists have used Knudsen cells for the purpose of evaporating the atoms under study. Mössbauer experiments, however, require 10^2-10^3 times the number of atoms necessary for ESR experiments. Therefore it is more advantageous to evaporate the metal atoms out of an open-ended cylindrical crucible (for example an alumina crucible 18 mm long and 3 mm ID as it has been used in most of the experiments reported here). The evaporation rate also is needed because of the above mentioned impurity effects: the high speed diffusion pumping system which has been used in the experiments reported here reduces the pressure to $2 \cdot 10^{-7}$ torr with the liquid nitrogen cold trap and to $\sim 8 \cdot 10^{-8}$ torr with liquid helium in the cryostat; this low pressure is absolutely necessary since even under these conditions and with the high evaporation rate the residual gas in the cryostat (N_2, O_2, etc.) will accrete on the absorber surface at a rate which is $\sim$5-10% that for the metal atoms in a typical run (metal deposition rate $\sim 10^{14}$ atoms s^{-1} cm^{-2}). This means that for lower evaporation rates, as is usually the case for Knudsen cells, an ultra high vacuum system with a residual vapor pressure of $< 10^{-9}$ torr is necessary for Mössbauer RGMI experiments.

The high metal atom deposition rate needed in Mössbauer RGMI experiments limits the RG deposition rate to at least 5-10 atomic layers/s. Lower matrix gas deposition rates give metal atom concentrations that are too high and consequently aggregate formation. This leads to additional lines in the Mössbauer spectra. Matrix gas deposition rates higher than 5-10 atomic layers/s give Mössbauer line broadening due to ill-defined environments (for example due to non-cubic lattice sites) /McNab et al., 1971a/.

2.3 Specific Problems in Electron Spin Resonance Experiments

Due to the low melting point, experiments with rare gas crystals involve, of course, low temperatures. To avoid clustering of the RGMI impurities by diffusion processes temperatures around the boiling point of liquid helium are thought to be low enough for these ESR experiments. Using low temperatures in an experiment is now almost trivial but ESR experiments also need a magnet with a limited pole gap and a microwave cavity with limited access to its interior. Thus ESR experiments on RGMI species are slightly hampered by geometric and temperature problems.

2.3.1 Sample Preparation

For almost all ESR experiments on RGMI atoms and ions samples condensed on a cooled substrate have been used. Owing to the low thermal conductivity of the solidified rare gases and the latent heat of solidification the thickness of a deposited sample is limited; nominally the thickness is about 1 mm. For the same reason a high thermal conductivity of the substrate is desirable. Depending on the experimental setup several materials are eligible.

In the work of Gordy and his group (see for example /Bouldin and Gordy, 1964/) the samples are deposited directly on the inside of the metallic wall of the microwave cavity. A cavity at 4 K poses, however, several problems such as the cooling of a relatively large mass, thermal drift of the resonance frequency, etc., so that most experiments used a room temperature cavity. Here the cooled substrate with the sample is inserted in the cavity. To avoid any disturbance of the ESR due to the substrate, the substrate has to show, besides its good thermal conductivity, a low electrical conductivity, low dielectric losses and a negligible number of paramagnetic centers. Synthetic sapphire or quartz is preferable for this purpose. Either discs with diameters up to 25 mm or rods with a diameter of several mm are used. These are rotated during deposition in order to cover homogeneously a larger area of the substrate.

Whereas the deposited samples usually seem to be polycrystalline and the trapping site of cubic or octahedral symmetry, Kasai and McLeod /1971/ observed distortions of the trapping sites along the direction which the matrix was grown on a sapphire substrate. To avoid undesirable interactions of the impurities with the substrate and the surface of the sample it is a well established practice to first condense a thin layer of pure matrix material and then anneal this film before the doped sample is deposited. Finally the sample is coated with pure matrix material. However, the growth direction may still influence the properties of the trapping sites in samples produced with a codeposition technique.

Holding the temperature as low as possible during and after the deposition of the sample would be favorable to avoid clustering of the impurities, whereas higher temperatures would favor a well annealed sample. For potassium atoms in an argon matrix Coufal et al. /1974a,1976/ showed that at temperatures below 20 K multiple trapping sites are observed that are annealed above 35 K. For the same matrix impurity system Goldsborough and Koehler /1964/ reported that some of the trapping sites can be only observed when the sample is exposed to visible or IR irradiation during deposition.

So far studies on multiple trapping sites indicate that it is advisable to deposit at temperatures as low as possible (for argon, temperatures well below 20 K are required), and to shield the growing matrix from visible and room temperature radiation during the deposition.

During the deposition, besides temperatures and possibly light irradiation, the gas flow rate is the most important parameter. Typical deposition rates are in the order of 10 atomic layers per second. *Experiments* on atoms or ions *isolated* in a rare gas matrix are *limited* of course, with respect to *the maximum allowable concentration of these impurities*. To avoid any impurity-impurity interactions, a ratio of one impurity atom or ion to a thousand rare gas atoms is generally considered to be an upper limit, lower concentration would be even preferable.

Using the deposition technique samples with roughly 10 cm^2 area and up to 1 mm thickness are deposited in several hours. Due to the above mentioned concentration limit samples produced by the deposition technique contain considerably less than $3 \cdot 10^{17}$ isolated atoms or ions.

In the experiments of Coufal et al. /1974b/ bulk samples with a diameter of 7.5 mm and 12 mm length were cut from a larger argon polycrystal grown by a crystal pulling technique. These pure argon crystals were subsequently doped with 0.2 ppm of potassium atoms by irradiation with thermal neutrons via the ^{40}Ar $(n,\beta^-)^{41}$K process. Due to the statistical generation of the impurity atoms they are homogeneously distributed within the sample. Because of limitations such as a thermal neutron flux of the order of $2 \cdot 10^{13}$ neutrons/s and irradiation times of the order of days, the maximum potassium concentrations in an argon crystal doped by neutron transmutation doping are less than 1 ppm. For safe handling of the sample (saturation activity in the above mentioned neutron flux 5 C/g) the sample volume should not exceed 0.5 cm^3. Due to these restrictions the samples produced by this technique contain, despite their bulk character, considerably fewer isolated atoms than typical deposition samples.

2.3.2 Experimental Setup

According to the properties of the samples the experimental setup for ESR experiments on RGMI atoms or ions is restrained. As shown in the previous paragraph limitations due to the samples are:

1. Low temperatures during sample preparation and measurement; the temperature should be well below 20 K.

2. A limited sample volume, normally less than 0.5 cm^3.

3. A small number of paramagnetic spins, usually much less than 10^{17} paramagnetic electron spins within a sample.

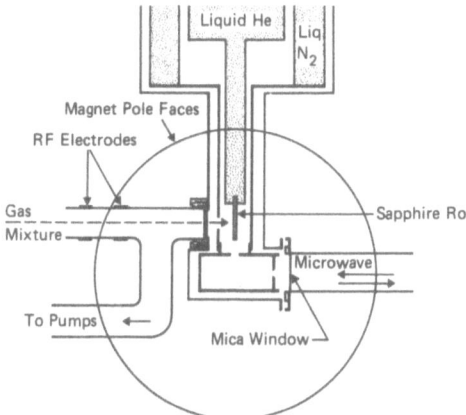

Fig.2.2. Experimental setup for ESR experiments on RGMI impurities /Adrian et al., 1962/

1. In many experiments sample preparation and resonance experiments are separated: most systems use a design after Duerig and Mador /1952/. As shown in Fig.2.2 a sapphire substrate is in direct thermal contact with liquid helium. After depositing the sample on the substrate, the inner part of the cryostat, the liquid helium dewar with the sapphire substrate and the sample can be brought from the deposition stage to the measuring position by lowering substrate and sample in the microwave cavity and slipping the complete assembly between the pole faces of the electromagnet. This setup has been used for most of the experiments reported here, but this basic concept had to be modified to fit the special conditions of the various experiments. In the experiments of Gordy and his group (see for example Rexroad and Gordy /1962/) the liquid helium flask, containing waveguide and a cavity with the sample deposited directly on its wall was inserted in a kilocurie ^{60}Co γ ray source to produce the paramagnetic impurities. In some of the irradiation experiments /Wall et al., 1959a,b; Coufal et al., 1974a,b/ special transfer equipment was used to transfer the sample from an irradiation cryostat within a γ source or a reactor into the microwave cavity of the ESR spectrometer without warming up the sample.

2. In spite of the small sample volume, in order to obtain a good filling factor for the microwave cavity, a small cavity that has a high microwave frequency seems favorable. Up to now, all experiments but Gordy's have been performed using X-band equipment in the 9 GHz range. Only Gordy's group uses a K-band spectrometer at 23 to 25 GHz. Several reasons may cause this preference for lower frequencies. X-band equipment is available in most laboratories and building attachments for the cavity is much easier to a bulky X-band workhorse than for tiny K-band equipment. In addition, proper selection of the cavity (mostly rectangular TE012 or cylindrical TE011 cavities) and optimum location of the sample at the maximum of the magnetic component of the microwave field are indispensable to have a reasonable filling factor especially at X-band frequencies.

3. Presently, most ESR spectrometers have a sensitivity in the order of $2 \cdot 10^{13}$ free spins per one Gauss (= 10^{-4} Tesla) linewidth at a signal to noise ratio of one. For a reasonable signal at a linewidth of several Gauss, which is typical for RGMI atoms or ions, at least $5 \cdot 10^{14}$ atoms or ions on identical trapping sites are necessary. Compared with the available number of spins even multiple trapping sites would not hurt.

But as with most low temperature ESR experiments, also in ESR on RGMI species, saturation becomes a problem. Spin-lattice relaxation times of S-state impurities are long at low temperatures especially when the impurities are only weakly coupled to a rare gas lattice. To avoid the saturation of the ESR transition only very low microwave power levels are permissible, usually in the μW range or even lower. So the main problem for ESR spectroscopy of RGMI atoms or ions is, besides sample preparation and handling, the access to an ESR spectrometer that is stable at these power levels. For most of the experiments reported here, ESR spectrometers with superheterodyne detection have been used, mostly modified Varian V-4500 equipment, but presently a wide scope of suitable spectrometers is commercially available.

3. Nuclear Gamma Resonance Experiments

3.1 Experiments on RGMI Atoms

3.1.1 ^{119}Sn

The Mössbauer γ transition in ^{119}Sn occurs between the I = 1/2 ground state and the I = 3/2 excited nuclear state; the γ transition energy is E_γ = 23.87 keV and the lifetime of the excited nuclear state is τ_γ = 17.75 ns. Tin has the atomic ground state

configuration $[Kr]4d^{10}5s^25p^2\ {}^3P_0$, i.e., is a non-magnetic (J = 0) atom. The Mössbauer absorption spectrum of RGMI ^{119}Sn atoms sitting on lattice sites with cubic point
symmetry (substitutional or octahedral interstitial sites), therefore, should exhibit a single resonance line whose position is determined by the electron density $\rho(0)$
at the ^{119}Sn nucleus. This is confirmed by Mössbauer absorption experiments on ^{119}Sn
atoms isolated in argon, krypton and xenon matrices (tin atomic concentration $\leq$ 1.0%)
/Micklitz and Barrett, 1972c/. All spectra show an unbroadened resonance line (see
for example Fig.3.1) with an IS of (+ 3.21 ± 0.01) mm/s relative to $BaSnO_3$ at 300 K independent of the RG matrix used. The IS of this resonance gives one calibration point
for the ^{119}Sn IS versus $\rho(0)$ scheme (see Sect.3.3.1).

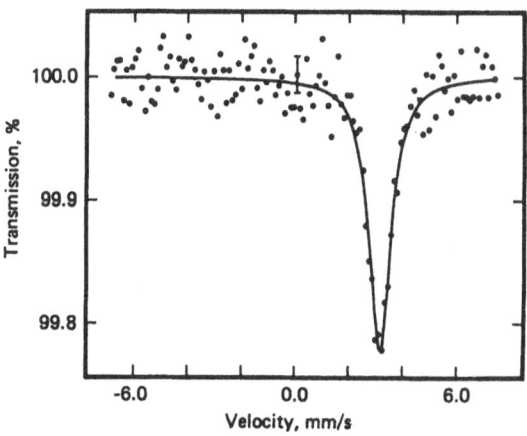

Fig.3.1. Mössbauer absorption
spectrum of ^{119}Sn atoms in a
xenon matrix at 4.4 K; Sn atom
concentration ~1%, Doppler velocity relative to ^{119m}Sn in
$BaSnO_3$ /Micklitz and Barrett,
1972c/

3.1.2 ^{57}Fe

The Mössbauer γ transition in ^{57}Fe between the I = 1/2 ground state and I = 3/2 excited nuclear state has an energy of E_γ = 14.41 keV; the lifetime of the I = 3/2
state is τ_γ = 97.7 ns.

Iron with the atomic ground state configuration $[Ar]3d^64s^2\ {}^5D_4$ is a magnetic atom
with a magnetic hf coupling constant A_0 = 38.1 MHz /Childs and Goodman, 1966/. Optical experiments with RGMI Fe atoms have shown /Mann and Broida, 1971; Micklitz
and Barrett, 1971/ that the crystal field splitting of the excited states is small
in comparison to the spin-orbit coupling, i.e., J is still a good quantum number for
RGMI-Fe. The cubic crystal field will split the $^5D_4$ ground state in one singlet, one
doublet and two triplets. The ground state of the system can be either a singlet (non-
magnetic) or a triplet (magnetic) depending on the sign of the crystal field interaction.

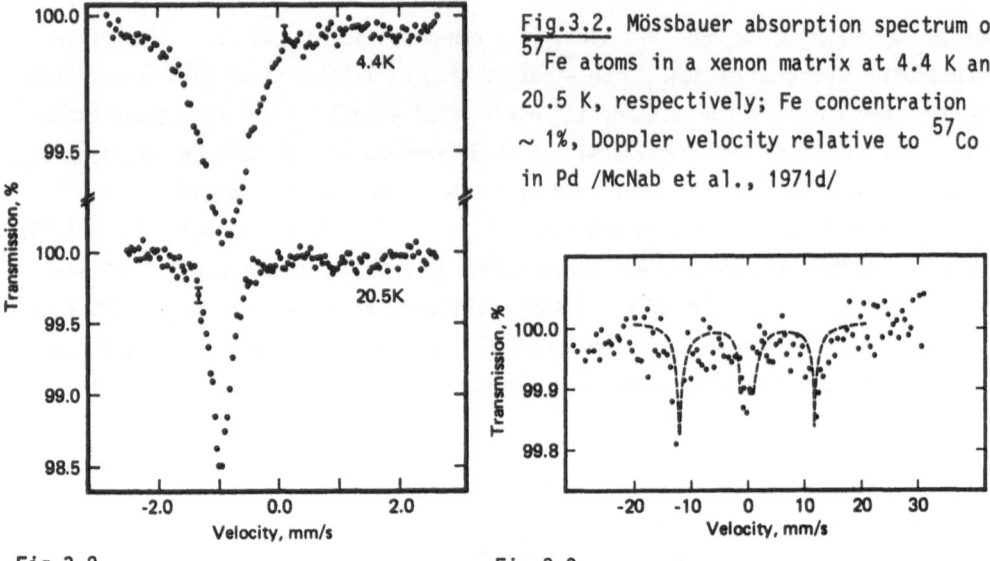

Fig.3.2. Mössbauer absorption spectrum of ^{57}Fe atoms in a xenon matrix at 4.4 K and 20.5 K, respectively; Fe concentration ~ 1%, Doppler velocity relative to ^{57}Co in Pd /McNab et al., 1971d/

Fig.3.2   Fig.3.3

Fig.3.3. Mössbauer absorption of ^{57}Fe atoms in a xenon matrix at 4.4 K in an external magnetic field of 2.8 T. The dashed line is a theoretical spectrum with relaxation times as given by Montano et al. /1974/

The Mössbauer absorption spectra of ^{57}Fe atoms isolated in argon, krypton and xenon matrices (iron atomic concentration < 1%) /McNab and Barrett, 1971; McNab et al., 1971/ show a single, strongly broadened resonance line, whose linewidth is temperature dependent (Fig.3.2). The IS of this resonance line is -(0.75 ± 0.03) mm/s relative to iron metal at 300 K independent of the RG matrix used. The temperature dependence of the resonance linewidth is linear with respect to 1/T (T = matrix temperature) and completely reversible. This led to the conclusion that the Fe atoms in the RG matrix are in a magnetic ground state, i.e., triplet. The magnetic hf interaction is strongly reduced by a fast paramagnetic relaxation (relaxation time, $\tau_{Rel} \ll \tau_{Larmor}$, nuclear Larmor precession time, via the direct phonon process (see also Sect.3.3.4). This was confirmed by later experiments with RGMI ^{57}Fe atoms in an external magnetic field B_{ext} /Montano et al., 1974,1975/. The Mössbauer absorption spectrum of ^{57}Fe atoms in a xenon matrix with B_{ext} = 2.8 T is shown in Fig. 3.3. Due to the large increase in τ_{Rel} caused by the external field a completely split magnetic hf pattern is observed ($\tau_{Rel} > \tau_{Larmor}$). The magnitude of the splitting corresponds to an effective magnetic hf field at the ^{57}Fe nucleus of B_{eff} = 70 ± 1.5 T (external field included). The magnetic hf fields at the ^{57}Fe nucleus of RGMI ^{57}Fe atoms are the largest internal fields observed so far at the ^{57}Fe nucleus. The magnitude of B_{eff} is that which one would expect for the triplet

$\{|T_2 1> = \frac{1}{\sqrt{8}} (\sqrt{7}|43> - |4-1>); \; |T_2 0> = \frac{1}{\sqrt{2}}(|42> - |4-2>); \; |T_2 -1> = \frac{1}{\sqrt{8}}(\sqrt{7}|4-3> - |41>)\}$

/Griffith, 1964/ being the ground state, if (i) the hyperfine coupling constant A of the free atom /Childs and Goodman, 1966/ is unchanged by the RG solid (see Sect. 3.3.2) and (ii) the crystal field splittings are $> kT$ (matrix temperature $T \sim 5$ K), i.e., only the triplet ground state is occupied. The fact that the triplet is the ground state means that the cubic crystal field parameter Δ for $Fe(3d^6 4s^2 \; ^5D)$ in RG solids is positive. The IS of the resonance line is determined by the electron density $\rho(0)$ at the ^{57}Fe nucleus of the $Fe(3d^6 4s^2 \; ^5D)$ configuration and is one important calibration point in the IS versus $\rho(0)$ scheme for ^{57}Fe (see Sect.3.3.1).

3.1.3 ^{151}Eu

The Mössbauer γ transition in ^{151}Eu occurs between the $I = 5/2$ ground state and the $I = 7/2$ excited nuclear state; the γ-transition energy is $E_\gamma = 21.64$ keV, the lifetime of the $I = 7/2$ state is $\tau_\gamma = 9.7$ ns. Europium has the atomic ground state configuration $[Xe]4f^7 6s^2 \; ^8S_{7/2}$, i.e., is a magnetic atom with a magnetic hf coupling constant $A_0 = -20.05$ MHz /Sanders et al., 1960/. The Eu $^8S_{7/2}$ ground state will split in a cubic crystal field due to the presence of a small admixture of non-S-excited states ($^6P_{7/2}$ and $^6D_{7/2}$) to the $^8S_{7/2}$ ground state. For $Eu^{2+}(4f^7 \; ^8S_{7/2})$ in an ionic crystal this admixture gives rise to a cubic crystal field splitting of the order 10^{-1} cm^{-1} /Baker et al., 1958/. For $Eu(4f^7 6s^2 \; ^8S_{7/2})$ in RG matrices this crystal field splitting Δ_{CF} may be even smaller due to the smaller crystal field strength in RG solids compared to that in ionic crystals, i.e., it is expected that Δ_{CF} of the $^8S_{7/2}$ state for Eu in RG matrices is $<< kT$ for matrix temperatures of $T \sim 5$ K. This is confirmed by optical emission experiments on RGMI-Eu atoms ($\Delta_{CF} << 10$ cm^{-1}) /Jakob et al., 1977/; optical absorption experiments on the same system show that the Eu atoms occupy non-cubic lattice sites in the RG solid /Jakob et al., 1976/.

In order to discuss the expected Mössbauer hf spectrum of RGMI-^{151}Eu the magnitude of Δ_{CF} with respect to the magnetic hf splitting Δ_{mhf} is important. (Δ_{mhf} is of the order of 10^{-3} cm^{-1}). For this reason we consider two extreme possibilities /Litterst et al., 1976/:

(i) $\Delta_{CF} << \Delta_{mhf}$. This is the "free atom" limit and the magnetic hf interaction is described by $\hat{H}_{hf}^{M1} = A \cdot \hat{I} \cdot \hat{J}$, J is the total angular momentum of the electronic ground state (J = 7/2).

(ii) $\Delta_{CF} >> \Delta_{mhf}$. In this case the magnetic hf interaction can be described by the independent interaction of the effective spins ($S_{eff} = 1/2$) of the four Kramer's doublets (degeneracy completely lifted due to low symmetry of Eu lattice site) with the nuclear spin I

$$\hat{H}_{hf}^{M_1} = \sum_{i=1}^{4} \frac{A}{g_J} (g_{eff})_i \, \hat{I} \cdot \hat{S}_{eff} \tag{3.1}$$

$g_J = 2$ is the g factor of the free Eu atom, $(g_{eff})_i$ are the unknown effective g factors of the four Kramer's doublets.

Figure 3.4 shows the measured absorption spectrum of ^{151}Eu atoms isolated in an argon matrix at 4.2 K /Litterst et al., 1976/. The solid curves in Figs.3.4a,b are different fits to the measured spectrum. In Fig.3.4a, the solid curve represents the "free atom" hf spectrum [case (i), $\Delta_{CF} \ll \Delta_{mhf}$] with the IS as the only free fitting parameter and assuming minimum linewidth ($\Gamma = 3.4$ mm/s) as given by the ^{151}Sm$_2$O$_3$ source for each hf component. The difference between this theoretical "free atom" and the measured hf spectrum may be caused by relaxation effects or by a small quadrupole interaction or by a small anisotropy in A due to the non-cubic lattice site of the Eu atom. This can be seen in Fig.3.4b where the solid curve represents a least squares fit for case (i). However, a small axial perturbation A_z is included in addition to the isotropic A(A_z/A = 0.04). For case (ii) ($\Delta_{CF} \gg \Delta_{mhf}$) no reliable fit to the measured absorption spectrum was possible due to the high number of free fit parameters (four unknown g factors).

The IS of the absorption spectrum given by its center of gravity is -(5.80 ± 0.05) mm/s relative to ^{151}Sm$_2$O$_3$ at 300 K. The consequences of this IS value with respect to the IS calibration for ^{151}Eu is discussed in Sect.3.3.1. The Mössbauer spectrum of ^{151}Eu atoms in an argon matrix has been remeasured by Montano /1982/. A slightly different IS value of -(6.1 ± 0.1) mm/s relative to ^{151}SmF$_3$ at 300 K is obtained. This difference, however, has no significance for the IS calibration as given in Sect.3.3.1.

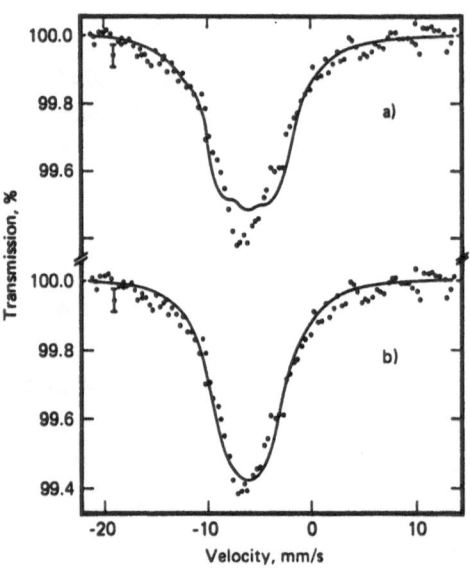

Fig.3.4. Mössbauer absorption spectrum of ^{151}Eu atoms in an argon matrix at 6 K; Eu concentration ~0.5%, Doppler velocity relative to ^{151}Sm$_2$O$_3$ at 80 K. The solid curve in a) is the simulated hyperfine spectrum of a free ^{151}Eu atom. The solid curve in b) is a least squares fit of the measured spectrum under the assumption that a small axial anisotropy A_z = 0.6 MHz is present in addition to the isotropic hyperfine coupling tensor A = -15.0 MHz /Litterst et al., 1976/

3.1.4 ^{125}Te

The Mössbauer γ transition in ^{125}Te between the I = 1/2 ground state and I = 3/2 excited nuclear state has the highest transition energy (E_γ = 35.46 keV, τ_γ = 1.56 ns) and recoil energy $E_R = E_\gamma^2/2Mc^2$ (M = nuclear mass) studied so far in RGMI experiments. This high recoil energy demands a cooled Cu-^{125}Sb source due to the low recoilless fraction (Lamb-Mössbauer factor, f) at room temperature (see Sect.3.3.3). Tellurium has the atomic ground state configuration $[Kr]5s^25p^4 \, ^3P_2$. Since tellurium evaporates out of Knudsen cells or open ended crucibles only as dimer Te$_2$ the procedure described in Sect.2.1.1 has been used to obtain RGMI ^{125}Te atoms. The Mössbauer absorption spectrum of ^{125}Te atoms in an argon matrix obtained after photodissociation of argon matrix isolated ^{125}TeH$_2$ by UV radiation (2500-3100 Å) from a xenon high pressure lamp is shown in Fig.3.5 /Montano et al., 1976/. The observed quadrupole splitting of 9.0 ± 0.02 mm/s produced by an electric field gradient at the ^{125}Te nucleus is probably caused by the fact that after photodissociation of ^{125}TeH$_2$ the hydrogen leaves the ^{125}Te atom in an unrelaxed, distorted surrounding. The IS of RGMI ^{125}Te is -0.1 ± 0.2 mm/s relative to Cu-^{125}Sb and gives one calibration point for the ^{125}Te IS versus $\rho(0)$ scheme (see Sect.3.3.1).

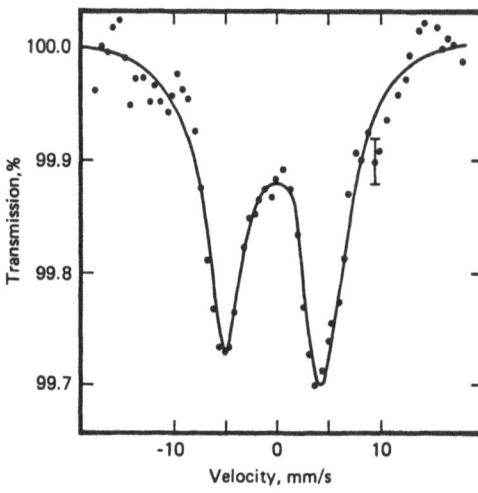

Fig.3.5. Mössbauer absorption spectrum of ^{125}Te atoms in an argon matrix at 4.2 K; Te concentration ~1.6%, Doppler velocity relative to ^{125}Sb in Cu at 80 K /Montano et al., 1976/

3.2 NGR Experiments on RGMI Ions

3.2.1 ^{57}Fe$^+$(3d^{6}4s^{6}D)

The generation and stabilization of the ^{57}Fe$^+$(3d^{6}4s^{6}D) ions was accomplished by the method described in Sect.2.1.2, i.e., by UV irradiation of a xenon matrix containing ~0.5% ^{57}Fe as an electron donor and ~1% HI as electron acceptor. The ^{57}Fe-doped Xe(HI)

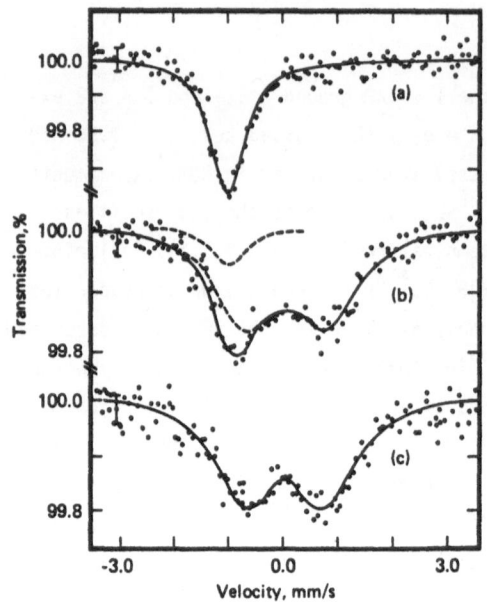

Fig.3.6. Mössbauer absorption spectra of ^{57}Fe atoms in a xenon matrix at 4.5 K; Fe concentration ~0.65%, Doppler velocity relative to ^{57}Co in Cu at 300 K. (a) Spectrum before UV irradiation, (b) spectrum taken after ~5 h of UV irradiation, (c) spectrum taken after ~15 h of UV irradiation /Micklitz and Litterst, 1974/

matrix was obtained by evaporating metallic iron (~80% enriched ^{57}Fe) out of an alumina crucible and mixing the iron atomic beam with a stream of Xe gas containing ~1% HI. This mixture was condensed on a liquid helium cooled beryllium disk. The amount of ^{57}Fe in the matrix was ~50 µg/cm^2 or ~5·10^{17} ^{57}Fe atoms/cm^2. For the UV irradiation of the matrix a 450 W high pressure xenon lamp together with a special filter combination, transparent for wavelengths from 2500-3000 Å was used. The UV photon flux was ~5·10^{15} s^{-1} cm^{-1} at the matrix.

Figure 3.6a shows the Mössbauer absorption spectrum of the ^{57}Fe-doped Xe(HI) matrix before UV irradiation. The spectrum shows a single resonance line with an IS of -0.74 ± 0.02 mm/s with respect to iron metal at 300 K. This spectrum is identical with that for ^{57}Fe atoms in a xenon matrix without HI admixture (see Sect.3.1.2). The Mössbauer spectra taken after ~5 h and ~15 h of UV irradiation are shown in Figs. 3.6a,b, respectively. After ~15 h of UV irradiation this resonance line is completely replaced by a quadrupole doublet with an IS of +0.26 ± 0.03 mm/s against iron metal at 300 K and a splitting of ΔE_Q = 1.39 ± 0.03 mm/s.

The quadrupole doublet in Figs.3.6a,b is interpreted as the resonance of the charged $Fe^+(3d^64s^6D)$ state. The quadrupole interaction is caused by the electric field gradient (EFG) of the $Fe^+(^6D)$ level split in the axial electric field of a neighbouring I^-. The size of the quadrupole splitting ΔE_Q varies from experiment to experiment depending on the iron and HI concentrations, i.e., depending on the average distance Fe^+-I^- /Montano et al., 1978/. The distribution of the Fe^+-I^- distances results in a variation of the $Fe^+(^6D)$ level splittings, and therefore in a distribution of the EFG at the iron nuclei. This is reflected in the large linewidth, $\Gamma = 1.42 \pm 0.02$ mm/s, of the quadrupole doublet.

The determination of the IS for the $^{57}Fe^{+}(3d^{6}4s^{6}D)$ resonance was an important step forward with respect to a reliable IS calibration for ^{57}Fe; with its help some of the existing discrepancies in the different IS calibration schemes for ^{57}Fe /Duff, 1976/ have been solved /Micklitz and Litterst, 1974/.

The hyperfine coupling constant A for RGMI-$Fe^{+}(3d^{6}4s^{6}D)$ could be determined in further experiments with an applied external magnetic field /Montano et al., 1978/. There exists no measured value for A of the free Fe^{+} ion. The detailed analysis of the magnitude of ΔE_Q as well as A (or B_{eff}) obtained from these RGMI experiments in connection with free iron ion spin polarized Hartree-Fock calculations showed une-quivocally that the crystal field ground state of Fe^{+} in xenon is a Kramer's doublet with the following eigenfunctions (axial field parameter $|\delta| \gg$ cubic crystal field parameter Δ):

$$|\psi_1\rangle = -\frac{\sqrt{7}}{4}\left|\frac{9}{2}\frac{9}{2}\right\rangle + \frac{\sqrt{2}}{4}\left|\frac{9}{2}\frac{1}{2}\right\rangle + \frac{\sqrt{7}}{4}\left|\frac{9}{2}-\frac{7}{2}\right\rangle \tag{3.2}$$

$$|\psi_2\rangle = \frac{\sqrt{7}}{4}\left|\frac{9}{2}-\frac{9}{2}\right\rangle - \frac{\sqrt{2}}{4}\left|\frac{9}{2}-\frac{1}{2}\right\rangle + \frac{\sqrt{7}}{4}\left|\frac{9}{2}-\frac{7}{2}\right\rangle \quad . \tag{3.3}$$

The sign of δ was found to be negative and that of Δ to be positive, i.e., Δ has the same sign as the cubic crystal field parameter for $Fe(3d^{6}4s^{2}\,^{5}D)$ in RG matrices (see Sects.3.1.2 and 3.3.5). A discussion of the measured A value for RGMI-$Fe^{+}(3d^{6}4s^{6}D)$ is given in Sect.3.3.2.

3.2.2 $^{57}Fe^{+*}(3d^{7})$

A charged state of the daughter atom ^{57}Fe due to the nuclear decay of ^{57}Co has been observed in solid xenon /Micklitz and Barrett, 1972b/. The ^{57}Co in solid xenon source was obtained by evaporating a source consisting of 1.3 mC ^{57}Co in Fe metal (90% en-riched ^{57}Fe). In this way a $^{57}Fe/^{57}Co$-doped xenon matrix with an iron atom concentra-tion of $\sim 1\%$ and ~ 10 μmC ^{57}Co was produced. This $^{57}Fe/^{57}Co$-doped xenon matrix was then used as (i) a ^{57}Co in solid xenon source and (ii) a ^{57}Fe in solid xenon absorber. The corresponding Mössbauer spectra are shown in Figs.3.7a,b,c, respectively. The Möss-bauer absorption spectrum (Fig.3.7c) is identical with that reported for ^{57}Fe doped xenon matrices (see Sect.3.1.2), i.e., the stable ^{57}Fe atoms are only in one state. The Mössbauer emission spectra (Figs.3.7a,b), however, show the ^{57}Fe atoms produced by the electron capture decay of ^{57}Co in two different states. One of these states is the stable Fe state with the atomic configuration $3d^{6}4s^{2}\,^{5}D$, the second state with the much more positive IS has a much lower electron density at the ^{57}Fe nuc-

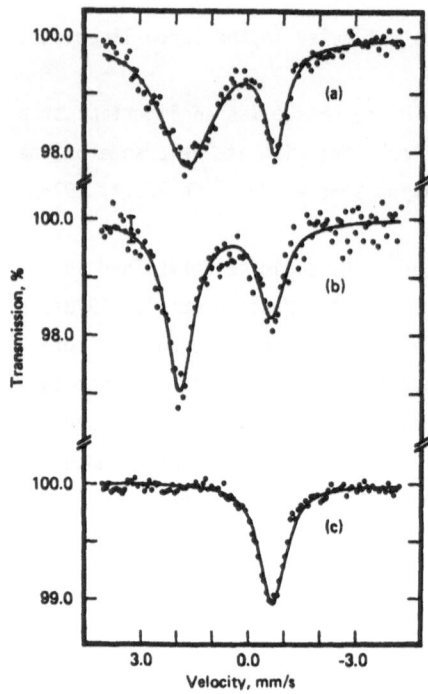

Fig.3.7. Mössbauer emission (a,b) and absorption (c) spectra of a $^{57}Fe/^{57}Co$ doped xenon matrix; Fe concentration ~1%, 10 µC ^{57}Co, matrix temperatures 4.4 K and 27 K, Doppler velocity relative to a K_4 ^{57}Fe $(CN)_6 \cdot 3H_2O$ absorber /Micklitz and Barrett, 1972b/

leus. All reported ^{57}Fe IS values lie essentially between the two resonances shown in Figs.3.7a,b. Following Micklitz and Barrett /1972b/ this second Fe state is an excited Fe^+ state with the atomic configuration $3d^7$. It is assumed that this Fe^+ state is preferentially occupied by a charge transfer process between Fe^{2+} and its xenon ligands $[Fe^{2+} + Xe \rightarrow Fe^{+*}(3d^7) + Xe^+]$. The temperature dependence of the recombination time, τ, for the process $Fe^+ \rightarrow Fe$ (the comparison between the line intensities in Figs.3.7a,b yields $\tau \sim T^{-1/2}$) indicates that this recombination takes place with thermal electrons. The IS of the $^{57}Fe(3d^7)$ resonance gives a third IS calibration point in the IS versus $\rho(0)$ scheme for ^{57}Fe (see Sect.3.3.1). Another Mössbauer source experiment on a ^{57}Co doped xenon matrix prepared by the ion implantation method (see Sect.2.3.1) was reported recently /Van Rossum et al., 1980/. The measured NGR spectrum and thus the IS of the $Fe^{+*}(3d^7)$ state is identical with that obtained by the codeposition of ^{57}Co and xenon /Micklitz and Barrett, 1972b/.

3.2.3 $^{119}Sn^+(4d^{10}5s^25p)$

An experiment with ^{119m}Sn in solid xenon has been performed that is similar to the source experiment described in the foregoing Sect.3.2.2 /Micklitz, 1977/. The Mössbauer emission spectrum shows a strongly broadened single resonance line which is attributed to $Sn^+(4d^{10}5s^25p^2P)$. It is assumed that the observed temperature dependence of the line width is caused by a temperature dependent paramagnetic relaxation of the magnetic $Sn^+(^2P)$ state [similar to $Fe(3d^64s^2\ ^5D)$, Sect.3.1.2]. The IS of the resonance is 3.34 ± 0.04 mm/s relative to $BaSnO_3$ at 300 K, i.e., differs by $+0.13 \pm 0.06$ mm/s from the RGMI $Sn(4d^{10}5s^25p^2\ ^1S_0)$ resonance due to the difference in the 5p shielding effect in these two states. The relevance of this small difference with respect to the IS calibration for ^{119}Sn is discussed in Sect.3.3.1.

3. 3 Discussion of Experimental Results

3.3.1 Isomer Shift

The IS of RGMI atoms and ions is the most important information obtained from Mössbauer experiments with RGMI species. The reason for it is that, in the interpretation and understanding of IS in solids, one is normally required to use free atom or free ion wave functions which have been modified to represent the atomic states in a real crystal. This approximation procedure leads to the calculation of electron densities at the nuclear $\rho(0)$ which can easily result in large errors in the determination of $\Delta<r^2>$ values. Different IS versus $\rho(0)$ schemes, therefore, differ considerably, depending on the IS calibration points and the approximation procedure used /Duff, 1974/. Since the IS of RGMI atoms and ions are supposed to represent the IS of "al-

Table 3.1. IS values of RGMI atoms and ions

	IS [mm/s]*
$^{57}Fe(3d^64s^2\ ^5D)$	-0.75 ± 0.03
$^{57}Fe^+(3d^64s^6D)$	$+0.26 \pm 0.03$
$^{57}Fe^{+*}(3d^7)$	$+1.77 \pm 0.08$
$^{119}Sn(4d^{10}5s^25p^2\ ^1S_0)$	$+3.21 \pm 0.01$
$^{119}Sn^+(4d^{10}5s^25p)$	$+3.34 \pm 0.05$
$^{125}Te(4d^{10}5s^25p^4\ ^3P)$	-0.20 ± 0.20
$^{151}Eu(4f^76s^2\ ^8S_{7/2})$	-5.80 ± 0.05

*IS values are given for a matrix temperature of 4.2 K with respect to the following Mössbauer absorbers or sources, respectively: Fe metal at 300 K for ^{57}Fe; Ba $^{119m}SnO_3$ at 300 K for ^{119}Sn; Cu ^{125}Sb at 70 K for ^{125}Te; $^{151}Sm_2O_3$ at 300 K for ^{151}Eu

most free" atoms (ions) such approximate procedures are not necessary in the calculation of $\rho(0)$ for RGMI atoms (ions) and the IS calibration obtained with such "calibration points" from RGMI experiments should be as reliable as the self-consistent
field (SCF) calculations of $\rho(0)$ for the different atomic (ionic) configurations.
Table 3.1 gives a survey of all measured IS of RGMI atoms and ions: $^{57}Fe(3d^64s^2\,^5D)$
has been studied in the RG matrices Ar, Kr and Xe and in the following other inert
matrices: N_2 /Micklitz and Barrett, 1972a/, CH_4 and CO_2 /Klee et al., 1974/. The IS
of the ^{57}Fe resonance is independent of the matrix material for all these inert matrices (experimental errors in IS are ±0.03 mm/s). The same was found for ^{119}Sn in
the matrices Ar, Kr, Xe and N_2 /Micklitz and Barrett, 1972a; Bos and Howe, 1974/.
This fact gives strong support for the assumption of "almost free" atoms (ions). It
shows that the influence of the RG solid on $\rho(0)$ of the RGMI atoms (ions) is negligible small. For example, a change in the IS for ^{57}Fe of ±0.03 mm/s corresponds to
a change of $\sim \pm 2\%$ in $\rho(0)_{4s2}$ [here $\rho(0)$ is the difference in the electron density
at the iron nucleus between the electron configuration given as the subscript and
the $Fe^{2+}(3d^6)$ configuration]. From this point of view the SCF cluster calculations
by Walch and Ellis /1973/ on $FeAr_{12}$, which give a change of $\sim -14\%$ in $\rho(0)_{4s2}$ at
the Fe nucleus due to the Ar ligands are questionable. The cancellation effect between the "overlap distortion" and "potential distortion" (Phillips theorem /Phil-
lips, 1959/) seems to be almost perfect in RG solids.

The three RGMI ^{57}Fe calibration points in Table 3.1 together with SCF calculations /Trautwein et al., 1975; Shenoy, 1974/ of $\rho(0)$ give a $\Delta<r^2>$ value for ^{57}Fe of
$-(14.1\pm 0.07)\cdot 10^{-3}$ fm^2 /Micklitz and Litterst, 1974/. This value is in agreement
with those obtained by Duff /1974/ from a critical re-evaluation of the existing IS
calibration attempts or by de Vries et al. /1975/, with the help of Dirac-Fock-Sla-
ter calculations. However, it is in disagreement with the low $\Delta<r^2>$ values given by
some other authors /Simanek and Wong, 1968; Pleiter and Kolk, 1971; Ruegsegger and
Kündig, 1972; Sharma and Sharma, 1972/. The IS differences between RGMI ^{57}Fe, $^{57}Fe^+$
and $^{57}Fe^{+*}$ give independent of $\Delta<r^2>$ the ratio $\rho(0)_{4s2}/\rho(0)_{4s} = 1.80\pm 0.15$. This
value is in good agreement with that obtained from optical isotope shift measure-
ments on *free* atoms for the 6s and 7s electron configurations: $\rho(0)_{ns2}/\rho(0)_{ns} =$
1.6 ± 0.1 for n = 6 /Kopfermann, 1956/ and n = 7 /Blaise and Steudel, 1968/. This
is further support for the concept of "almost free" atoms (ions) used in the analysis of the IS data for RGMI atoms (ions).

In the case of ^{119}Sn there is essentially only one IS calibration point, namely
RGMI $^{119}Sn(4d^{10}5s^25p^2\,^1S_0)$. The IS difference between the RGMI ^{119}Sn and the RGMI
$^{119}Sn^+(4d^{10}5s^25p)$ resonance (see Table 3.1) is of the same order or magnitude as possible RG matrix effects on $\rho(0)$ $[\Delta\rho(0)/\rho(0)_{5s2}\sim 4\%]$, i.e., the RGMI $^{119}Sn^+$ IS can not
be taken as a second IS calibration point. Also there is only one IS calibration

point for ^{125}Te (see Table 3.1). However, combining the IS of RGMI ^{119}Sn and of RGMI ^{125}Te($4d^{10}5s^25p^4$ ^{3}P) with the $\Delta<r^2>/<r^2>$ ratios /Ruby and Shenoy, 1969/ as obtained from the concept of isoelectronic compounds /Ruby et al., 1967/ one can give lower and upper limits for $\Delta<r^2>$ of ^{119}Sn as well as of ^{125}Te /Micklitz, 1977/:

$$^{119}\text{Sn: } 3.6\cdot10^{-3} \text{ fm}^2 \le \Delta<r^2> \le 5.6\cdot10^{-3} \text{ fm}^2$$

$$^{125}\text{Te: } 2.9\cdot10^{-3} \text{ fm}^2 \le \Delta<r^2> \le 4.5\cdot10^{-3} \text{ fm}^2 \quad .$$

The $\Delta<r^2>$ value for ^{151}Eu can be obtained independently of Mössbauer IS data from the analysis of μ mesonic data together with nuclear model calculations ($\Delta<r^2>$ = $18\cdot10^{-3}$ fm^2) /Meyer and Speth, 1972/. The IS for RGMI ^{151}Eu together with the extra-polated values of the IS for ^{151}Eu^{2+}($4f^7$) and ^{151}Eu^{3+}($4f^6$) /Gerth et al., 1968/ give $\Delta<r^2>$ = $22\cdot10^{-3}$ fm^2 /Litterst et al., 1976/, close to the value obtained from μ me-sonic data. The IS of RGMI ^{151}Eu, however, is in serious disagreement with the value predicted from the analysis of optical isotope shift data /Brix et al., 1964/. Ac-cording to this analysis $\rho(0)$ in Eu($4f^76s^2$) should be larger than $\rho(0)$ in Eu^{3+}($4f^6$), which means that the IS of ^{151}Eu($4f^76s^2$) should be more positive than that of ^{151}Eu^{3+}($4f^6$). Therefore, it has been concluded /Litterst et al., 1976/ that the as-sumption made in the optical isotope shift data analysis that shielding effects for similar configurations of Sm and Eu are not very different, is not justified. This is confirmed by relativistic SCF calculations of Coulthard /1973/.

3.3.2 Hyperfine Coupling Constant

The only RGMI Mössbauer experiments which give information about the magnetic hf field present at the nucleus of the RGMI atom (ion) are the experiments with RGMI ^{151}Eu($4f^66s^2$ $^8S_{7/2}$) and those with RGMI ^{57}Fe($3d^64s^2$ ^{5}D) and RGMI ^{57}Fe$^+$($3d^64s^6$D) in the presence of an external magnetic field (quasistatic condition, i.e., paramag-netic relaxation time much longer than nuclear lifetime).

The comparison of the measured RGMI ^{151}Eu Mössbauer spectrum with the hf spectrum of the free ^{151}Eu atom using the known hyperfine coupling constant A = -20.05 MHz /Sanders and Woodgate, 1960/ (see Fig.3.4a) indicates that A for the RGMI ^{151}Eu atom is essentially the same as that for the free ^{151}Eu atom (same overall splitting). The experimental data, however, do not allow one to decide if a small change in A of the order $\Delta A/A_0 < 10\%$ is present (this is the order of magnitude in $\Delta A/A_0$ which is usually observed in RGMI-ESR experiments, see Chap.4).

For RGMI ^{57}Fe and RGMI ^{57}Fe$^+$ in the presence of an external magnetic field the hf coupling constants A or B_{eff} [$A(g_{eff}/g)\cdot I\cdot S_{eff} = g_N\beta_N\cdot I\cdot B_{eff}$, see Chap.1] have been determined more accurately (see Table 3.2). The predicted values for H_{eff} have

Table 3.2. Comparison of measured and predicted values of the magnetic hyperfine field B_{eff}/T at the ^{57}Fe nucleus of RGMI-Fe and Fe+

	Measured Value	Predicted Value*
Fe($3d^6 4s^2\,^5D$) Ar	81 ± 1.5	68.8
in Xe	67 ± 1.5	
(cubic symmetry)		
Fe+($3d^6 4s^6D$) in Xe	$H_x = H_y = 70 \pm 1$	$H_x = H_y = 73.6$
(axial symmetry)	$H_z = 35 \pm 1$	$H_z = 36.8$

*The predicted value of B_{eff} has been obtained from free atom (ion) A_0 values assuming no mixing of excited crystal field states to the crystal field ground state, i.e., $\Delta_{CF} > \Delta_{Zeeman}$ at $B_{ext} = 3.0$ T, $\Delta_{CF} > kT$ at T = 4.2 K

been calculated from the free atom (ion) values and the effective g factors of the ground state wave functions. In the case of Fe the free atom value of A has been determined experimentally /Childs and Goodman, 1966/; for Fe+ there exists only a theoretical value for A of the free ion /Montano et al., 1978/. The crystal field ground state for Fe($3d^6 4s^2\,^5D$) in cubic symmetry has to be the triplet ($S_{eff} = 1$) given in Sect.3.1.2 with (g_{eff}/g) = 5/2. The crystal field ground state for Fe+($3d^6 4s^6D$) was determined by the condition that it has to fit both for the magnitude of the measured ΔE_Q and B_{eff}. This condition determines unequivocally the ground state wave function (Kramer's doublet, $S_{eff} = 1/2$) given in Sect.3.2.1 with an anisotropic g tensor (axial symmetry): $(g_{eff}/g)_x = (g_{eff}/g)_y = (g_{eff}/g)_\perp = 2$; $(g_{eff}/g)_z = (g_{eff}/g)_{||} = 1$ ($||$ and $\perp$ denote the directions $||$ and $\perp$ with respect to H_{ext}). The agreement between the measured and predicted B_{eff} values is excellent for Fe and Fe+ in solid xenon. Possible changes $\Delta B_{eff}(\Delta A)$ and $B_{eff}(A)$ due to the xenon matrix are < 5%. For Fe in solid argon there is a definite change in $B_{eff}(A)$ due to the argon matrix. It is assumed /Montano et al., 1976/ that this change is caused by the mixing of excited crystal field states with the ground state triplet in the presence of the external field. This means that the crystal field splitting Δ_{CF} is of the same order of magnitude as the Zeeman splitting of the ground state triplet $\Delta_{Zeeman} = g_{eff} \cdot \beta_e \cdot S_{eff} \cdot B_{ext}$, with $B_{ext} = 3.0$ T. Such a mixing seems to be negligible for Fe in solid xenon, i.e., $(\Delta_{CF})_{Xe} > (\Delta_{CF})_{Ar}$ for the $3d^6 4s\,^5D$ Fe ground state. An even larger value for B_{eff} has been observed for ^{57}Fe in a nitrogen matrix /Montano et al., 1976/; this is an indication that for Fe in solid nitrogen Δ_{CF} is even smaller than in solid argon, i.e., $(\Delta_{CF})_{N_2} < (\Delta_{CF})_{Ar} < (\Delta_{CF})_{Xe}$.

3.3.3 Lamb-Mössbauer Factor

The recoilless fraction, f, the so-called Lamb-Mössbauer factor of the nuclear γ resonance absorption, is determined by the nuclear recoil energy $E_R = E_\gamma^2/2Mc^2$ (E_γ: γ transition energy, M: mass of Mössbauer isotope), the phonon spectrum $N(\omega)$ of the solid and the absorber temperature T /Visscher, 1960; Wegener, 1965/:

$$-\ell nf = 2W = <x^2(T)>/\bar{x}^2 = \frac{2E_R}{h} \int_0^\infty d\omega \, \frac{N(\omega)}{\omega} \left(\frac{1}{2} + \frac{1}{e^{\hbar\omega/kT}-1} \right) \quad . \tag{3.4}$$

$<x^2(T)>$ is the mean square displacement of the Mössbauer nucleus at the absorber temperature T, $\bar{x}$ is the γ ray wavelength divided by 2π. If $N(\omega)$ is described in a Debye model $[N(\omega) \propto \omega^2]$ with an effective Debye or Mössbauer temperature θ_M, 2W can be written as /Mössbauer and Wiedemann, 1960/:

$$2W = \frac{3E_R}{k\theta_M} \left[\frac{1}{2} + 2\left(\frac{T}{\theta_M}\right)^2 \int_0^{\theta_M/T} \frac{x}{e^x-1} \, dx \right] \quad . \tag{3.5}$$

The measurement of the temperature dependence of f allows the determination of θ_M with help of the formulas (3.4,5). Table 3.3 gives the θ_M values for ^{57}Fe in the RG matrices Ar, Kr and Xe obtained in this way /McNab et al., 1971/ together with the

Table 3.3. Comparison of Debye-Waller temperatures θ_{DW}^h, as obtained from neutron scattering measurements on RG solids, "impurity" Debye temperatures θ^i and Mössbauer temperatures θ_M as probed by RGMI atoms and ions

		Ar	Kr	Xe
	$\theta_{DW}^h(K)$ [a]	83	64	56
^{57}Fe	$\theta_M(K)$	61 ± 4	56 ± 4	59 ± 4
	$\theta^i(K)$ [b]	69	77	83
^{57}Fe$^+$	$\theta_M(K)$			80 ± 5
	$\theta^i(K)$ [b]			83
^{151}Eu	$\theta_M(K)$	22 ± 4		
	$\theta^i(K)$ [b]	42		

[a] Powell and Dolling /1977/
[b] Assuming $\theta^i = \theta_{DW}^h (M_h/M_i)^{1/2}$ /Visscher, 1963/

corresponding Debye-Waller temperatures (at 10 K) of the RG solids ("host" Debye-Waller temperature, θ_{DW}^h, /Powell and Dolling, 1977/ and the effective Debye temperatures ("impurity" Debye temperature, θ^i) taking the different mass M_i of the impurity atom ^{57}Fe with respect to the mass M_h of the RG host atoms into account $[\theta^i = \theta_{DW}^h (M_h/M_i)^{1/2}]$ /Visscher, 1963/. A comparison between θ_M and θ^i shows that $\theta_M < \theta^i$ for all RG solids. This in turn implies that force constant changes occur and that a simple mass scaling is inadequate. An attempt has been made /Paul and Puri, 1982/ to calculate the recoilless fraction, f, for ^{57}Fe in Ar, Kr and Xe using experimental determined phonon frequency distribution functions /Fujii et al., 1974/; Lurie et al., 1974; Skalyo et al., 1974/ and correcting for anharmonicity. It is found that the ratio of the impurity-host to host-host coupling-force constants is changed in such a way, that the f values remain the same in these lattices. No localized modes are found due to the weak binding of the impurity to the host.

The measurement of the temperature dependence of *optical* absorption lines from iron atoms in RG matrices offers the possibility to determine θ_{DW}^i of such an iron atom. According to a theory of Lax /1952/ the square of the linewidth of such absorption lines should be in first approximation proportional to $<x^2(T)>$, i.e., proportional to 2W. Such a measurement has been done with an iron-doped krypton matrix /Micklitz and Barrett, 1971/. From the temperature dependence of $<x^2(T)>$ one gets in the Debye model $\theta_{DW}^i = 60 \pm 5$ K. This value is in good agreement with that determined from the Mössbauer effect ($\theta_M = 56 \pm 4$ K for ^{57}Fe in Kr, see Table 3.3).

For ^{151}Eu in solid argon the recoilless fraction, f, has been estimated from the amount of ^{151}Eu in the argon matrix and a comparison of the observed resonance absorption area of ^{151}Eu in solid argon with the resonance absorption area of a calibrated ^{151}Eu$_2$O$_3$ absorber. A value of $f(T = 6$ K$) = 0.18 \pm 0.04$ is obtained /Litterst et al., 1976/. This corresponds to $\theta_M = 22 \pm 4$ K ($\triangleq \hbar\omega_M = 15 \pm 3$ cm^{-1}) which is much lower than expected from the argon phonon spectrum: $\theta^i = (M_{Ar}/M_{Eu})^{1/2} \theta_{DW}^{Ar} = 42$ K. This very low value for f or θ_M respectively was satisfactory explained by the results of an optical absorption experiment with Eu atoms isolated in an argon matrix /Jakob et al., 1976/. This optical absorption spectrum due to $4f^7 6s^2 \, ^8S_{7/2} \rightarrow 4f^6 5d6s^2$ transitions shows zero-phonon lines accompanied by phonon-side bands caused by resonance modes in the phonon spectrum with $\hbar\omega_1 = 12 \pm 2$ cm^{-1} and $\hbar\omega_2 = 26 \pm 2$ cm^{-1} ($\hbar\omega_D^{Ar} = 67$ cm^{-1} /Batchelder et al., 1970/). Due to these low-frequency modes in $N(\omega)$ the mean square displacement $<x^2(T)>$ of the Eu atom in the argon matrix is strongly increased ($<x^2(T)> \sim \int_0^\infty d\omega N/\omega$ for $T << \theta_D$ and therefore f and θ_M are strongly reduced).

The observation that for ^{57}Fe in RG solids $\theta_M < \theta^i$ (see Table 3.3) leads to the supposition that such low-frequency modes might also be present in the phonon spectra of Fe doped RG solids.

The effective Debye temperature θ_M for $Fe^+(3d^64s^6D)$ in xenon as obtained from the temperature dependence of the resonance absorption area is $\theta_M = 80 \pm 5$ K /Montano et al., 1978/. This is in good agreement with the θ_M value measured for $Fe^+(3d^7)$ in xenon matrix: $\theta_M = 90 \pm 20$ K /Micklitz and Barrett, 1972b/. The higher value of θ_M for monovalent iron in comparison to that for neutral iron can be explained by the polarization of the RG nearest neighbor atoms that produce a stronger coupling of the iron ion with the matrix atoms.

3.3.4 Spin-lattice Relaxation

The linewidth of the ^{57}Fe resonance in RG matrices is determined by the paramagnetic spin-lattice relaxation time of the crystal field ground state (triplet). The observed temperature dependence of the resonance linewidth, which is completely reversible, gives information about the temperature dependence of the spin-lattice relaxation time T_1. The analysis yields $\Gamma = \Gamma_0 + \Delta\Gamma$, $\Delta\Gamma \sim T_1 \sim T^{-1}$ (Γ_0 = natural linewidth, T = RG matrix temperature /McNab et al., 1971/. This is an indication that the direct phonon process is dominant for the spin-lattice relaxation of Fe in RG matrices. The crystal field splitting of the $Fe(3d^64s^2{}^5D)$ state is smaller than $k\theta_D$ (θ_D = Debye temperature of RG solid, see Sect.3.1.2) thus allowing the direct phonon process for the spin-lattice relaxation mechanism. The experimental evidence for an unquenched orbital momentum (J good quantum number) together with the fact that the relaxation takes place between states which are not Kramer's doublets leads one to expect a very short spin-lattice relaxation time even at low temperatures /Orbach, 1962/. The observed relaxation time of $T_1 \sim 0.25$ ns at 1.45 K for ^{57}Fe in RG matrices, however, is much shorter than usually observed, for example, in Mössbauer experiments with rare earth salts where there are also unquenched orbital momenta and relaxation between non-Kramer's doublets /Hüfner et al., 1968/. This can be explained by the very low Debye frequencies of the RG solids compared to that of ionic crystals [θ_D(RG) $\sim$ $0.10\theta_D$ (ionic crystal)] and the fact that $T_1 \propto N(\omega)^{-1}$ $v^2 \propto \omega_D^5$ [$N(\delta)$ is the phonon density of states at $\omega = \delta$, where δ is the splitting of the states involved in the relaxation process, v is the velocity of sound]. The possible existence of low-frequency modes in the phonon spectra of Fe doped RG solids (see Sect.3.3.3) would give a further decrease in T_1.

3.3.5 Crystal Field Parameters

Optical absorption experiments on RGMI atoms or ions give directly information about the crystal field splitting Δ_{CF} of the *excited* atomic or ionic states, i.e., the crystal field parameters of the RG crystal field acting on these excited states. Optical emission experiments yield in principle the same information about the crystal field parameters relevant for the atomic (ionic) *ground* state. However, such experiments have been done only on RGMI alkali atoms /Belyaeva et al., 1969,1973; Micklitz and Luchner, 1974/ which do not show a ground state splitting (S state).

The analysis of the hf data obtained from Mössbauer absorption experiments on RGMI atoms (ions) give indirectly some information about Δ_{CF} in the atomic (ionic) *ground* state.

For example, the experiments on RGMI ^{57}Fe and xenon matrix isolated ^{57}Fe$^+$ with external magnetic field (see Sects.3.1.2 and 3.2.1) give unequivocally the crystal field ground state wave function and therefore the sign of the cubic crystal field parameter Δ. The sign of Δ is positive for RGMI Fe as well as for Fe$^+$($3d^6 4s^6$D) in a xenon matrix.

Further information is obtained about the order of magnitude of Δ_{CF}: the relatively good agreement between the measured and the predicted magnetic hf fields at the RGMI ^{57}Fe nucleus and the ^{57}Fe$^+$ nucleus in a xenon matrix shows that (i) J is still a good quantum number for Fe and Fe$^+$ in solid xenon ($\Delta_{CF} < \lambda \cdot \mathbf{L} \cdot \mathbf{S}$, the spin-orbit interaction) and (ii) no thermal occupation of excited crystal field states is present at a matrix temperature of ~4.2 K, i.e., 10 cm^{-1} $\lesssim \Delta_{CF} \lesssim$ 100 cm^{-1} for the Fe ground and the Fe$^+$ ground state in a xenon matrix. For ^{57}Fe in solid argon B$_{eff}$ is ~15% larger than expected (see Table 3.2); this is an indication that excited crystal field states are mixed into the ground state due to the Zeeman splitting (Δ_{Zeeman} ~10 cm^{-1} at H$_{ext}$ = 3 T) in the presence of an external field, i.e., Δ_{CF} is of the order 10 cm^{-1} for the Fe ground state in solid argon. Such a mixing seems not to be present in solid xenon, i.e., Δ_{CF}(Xe) > Δ_{CF}(Ar) for the Fe ground state.

The order of magnitude of Δ_{CF} for Eu in an argon matrix as obtained from the Mössbauer experiment is the following: $\Delta_{CF} \ll$ kT ~3 cm^{-1}; Δ_{CF} is eventually even of the order of Δ_{mhf} ~10^{-3} cm^{-1} (see Sect.3.3.3). The latter would not be too surprising since Eu has a $^8S_{7/2}$ ground state and Δ_{CF} is only caused by a small admixture of $^6P_{7/2}$ and $^6D_{7/2}$ excited states to the ground state /Baker et al., 1958/.

The fact that no quadrupolar broadening of the resonance lines for RGMI ^{57}Fe as well as for RGMI ^{119}Sn is observed shows that the lattice sites of these RGMI atoms have essentially cubic symmetry. However, this is not the case for Eu atoms in an argon matrix; this can be seen directly in the optical spectra of RGMI Eu atoms /Jakob et al., 1976/ but also indirectly in the Mössbauer absorption spectrum of argon isolated ^{151}Eu atoms (see Sect.3.1.3).

NGR spectroscopy on RGMI atoms and ions has proved to be a successful experimental method for the purpose of the Mössbauer IS calibration. This is well demonstrated for ^{57}Fe where three different electronic configurations of RGMI Fe and thus three ^{57}Fe-IS calibration points could be obtained.

Due to fast paramagnetic spin-lattice relaxation in RGMI atoms or ions with un-quenched orbital momentum, the magnetic hf interaction in RGMI atoms or ions cannot be measured in an ordinary RGMI-NGR experiment. The application of an external mag-netic field, however, increases the spin-lattice relaxation time which allows the determination of the magnetic hf coupling constant. For example, the hf coupling constant of ^{57}Fe^{+}($3d^{6}4s\ ^{6}$D), which has not been measured before, was determined by such an experiment.

Information about the lattice site symmetry of the impurity atom or ion and the crystal field parameters for the electronic ground state can be obtained indirect-ly from the analysis of the RGMI-NGR spectra.

Measurements of the temperature dependence of the NGR recoilless fraction give an indication of the existence of low-frequency resonant modes in the phonon spectra of doped RG solids.

4. ESR Experiments

Electron Spin Resonance (ESR) is both a very sensitive and a high resolution method for spectroscopy of paramagnetic impurities. It proved to be a powerful tool for elu-cidating the electronic structure of defects, atoms and molecules in crystals and so-lutions; color centers in alkali halide single crystals are a well known example. The deep understanding of these paramagnetic centers came from ESR experiments on single crystals.

ESR experiments on RGMI atoms or ions suffer from an inherent difficulty: samples are polycrystalline or even amorphous. In addition most experiments deal with atoms or ions with zero angular momentum. Therefore, from most of these experiments, only isotropic Zeeman parameters (g_{eff} factors) and isotropic hyperfine interaction con-stants (A_{eff} values) of orbitally non-degenerate (S state) species could be derived. Only in experiments of Ammeter and Schlosnagle /1973a,b/ on RGMI Al and Ga ($^{2}P_{1/2}$ state) and of Iwasaki et al. /1979/ on RGMI iodine atoms ($^{2}P_{3/2}$ state) the elements of the g_{eff} and A_{eff} tensors of the trapped species could be determined. In experi-

ments with S ground state impurities ESR has contributed very little beyond the identification of the RGMI species by its characteristic hyperfine pattern and a clear
distinction between different trapping sites (e.g. Gruen /1976/). Almost nothing
about the symmetry of these sites could be learned. Theoretical models on the matrix shift of S state impurities describe some trends correctly but are too approximate for a numerical comparison with experiments for a given cluster geometry. That
is why Ammeter and Schlosnagle's experiments and their theoretical understanding was
both an experimental and theoretical breakthrough. Besides the above mentioned ESR
experiments double resonance experiments on K and Rb atoms in argon matrices were
very effective in assigning different trapping sites of these alkali atoms, as identified by ESR, to the various observed optical absorption bands.

All ESR experiments on RGMI species with a S ground state symmetry are consistent
with the assumption of an essentially centro-symmetric trapping site. The parameters
A_{eff} and g_{eff} in the Hamiltonian of the trapped atom with a S ground state can then
be written as

$$A_{eff} = A = A_0 + \Delta A \tag{4.1}$$

$$g_{eff} = g = g_0 + \Delta g \tag{4.2}$$

where A_0 and g_0 are parameters of the free atom and ΔA and Δg are the deviations from
the free atom parameters as a result of the matrix environment.

In the following part of this section, ESR experiments on RGMI atoms and ions are
briefly reviewed. Numerical results for the relative hyperfine shifts $\Delta A/A_0$ and the g
factor shifts Δg are given. These data were compiled from the original papers using
free atomic values A_0 and g_0 from Kusch and Hughes /1959/ or Ayscough /1967/; when no
reliable data for g were available a free electron value of $g = 2.0023$ /Kusch and Hughes,
1959/ was used as a reference. For the conversion of the data, the following numerical
values for the Bohr magneton /Handbook of Chemistry and Physics, 1978/ were used

$$\mu_B = 5.788381 \cdot 10^{-5} \text{ eV T}^{-1}$$
$$= 1.3996108 \cdot 10^{10} \text{ s}^{-1} \text{ T}^{-1}$$
$$= 46.68598 \cdot 10^{-2} \text{ cm}^{-1} \text{ T}^{-1} .$$

The assignment of data to the various trapping sites may be sometimes rather arbitrary.
Nevertheless it may prove helpful in the comparison of various data of one dopant in
different matrices or a series of different impurities trapped in the same type of
rare gas matrix. For each impurity in every matrix the various trapping sites are num-
bered according to the relative hyperfine shift $\Delta A/A_0$, starting at the largest observed negative shift with number 1 and continuing with decreasing shift. During the evaluation of these data it became obvious that the sample preparation plays an important
role for the trapping. Therefore besides the spectroscopic data also the sample prepa-

ration technique leading to a trapping on a particular site is specified using the following abbreviations:

C Codeposition
P Photolysis of deposited sample
γ γ irradiation of deposited sample
n n irradiation of grown crystal

4.1 Paramagnetic Impurities with S Ground State Symmetry

Most RGMI impurities have, despite their different electron configuration, a S ground state symmetry. As mentioned above, therefore, only very little on the symmetry of their trapping sites is known. But due to the fact that samples with these impurities, i.e., hydrogen or the alkali-metals are relatively easy to prepare almost all results on RGMI single impurities come from S ground state species.

4.1.1 Atoms with $(ns)^1$ Electron Configuration

Hydrogen. Of all RGMI atoms with one unpaired S electron hydrogen is the impurity that has been observed in the largest number of ESR experiments, probably because H is a by-product in experiments on RGMI radicals. Some authors like Bouldin and Gordy /1964/ that studied the production of H from RGMI H_2 under γ irradiation were only interested in integral intensities of the H ESR lines. Therefore they and some other groups /e.g., Fischer et al., 1967; Kasai, 1968/ that were not mainly interested in the by-product H did not give data on the spin Hamiltonian. Nevertheless, sufficient data on H in different matrices is available (Table 4.1). Multiple trapping sites (see Fig.4.1) are evident for all matrices but Ne.

Only one experiment on Ne:H has been reported by Foner et al. /1960/ showing one single trapping site. This may be due to the close matching of the size of H and Ne atoms or to the relatively low melting point of the neon matrix which thermally anneals the sample during deposition or observation of the spectra. This may explain why H atoms when trapped in a Ne matrix are only in one trapping site (probably substitutional).

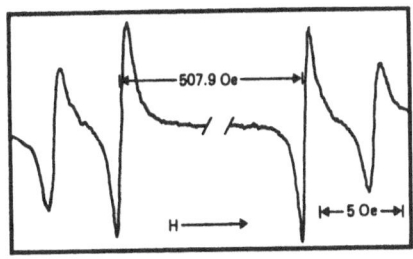

Fig.4.1. ESR spectrum of hydrogen atoms produced by photolysis of water in solid argon at 4.2 K. The spectrum demonstrates the existence of two different, 5 Oe shift, trapping sites for hydrogen atoms that are characterized by their hyperfine doublet with a splitting of 507.9 Oe /Adrian et al., 1962/

Table 4.1. Hyperfine and g factor shifts for RGMI hydrogen atoms on various trapping sites

Matrix	Site	$\frac{\Delta A}{A_0} \cdot 10^2$	$-\Delta g \cdot 10^3$	Sample Preparation	Author
Ne		0.43	0.19	P	a
Ar	1	-0.46	0.06	C	b
		-0.46	0.06	P	c
		-0.46	0.06	C	a
		-0.52		C	d
	2	-0.29	0.02	P	c
		-0.29	0.02	P	a
	3	1.15	0.65	P	c
		1.15	0.65	P	a
		1.15	1.18	Y	e
		1.09		C	d
Kr	1	-0.59	0.47	C	a
		-0.59	0.47	C	a
	2	0.47	2.59	P	a
		0.53	3.37	Y	e
Xe	1	-1.09	0.57	C	a
	2	-1.04	1.69	P	a
		-1.02	2.27	Y	e

a Foner et al. /1960/   c Cochran et al. /1959/   e Jackel et al. /1968/
b Jen et al. /1958/   d Goldsborough and Koehler /1964/

For Ar:H three well defined trapping sites are reported by several authors. Strik-
ing is the fact that sites 2 and 3 could be only observed in samples prepared by a
technique employing relatively high local energies like UV photolysis or γ irradiation. Site 3 has been observed also by Goldsborough and Koehler /1964/ in a codeposition experiment. But they report that this site could be observed only with one
sample and that they were unable to reproduce this particular result. According to
Cochran et al. /1959/ and confirmed by Foner et al. /1960/ site 1 is stable against
annealing up to 39 K, whereas site 2 becomes unstable at 23 K and site 3 disappears
above 12 K. Their analysis suggests that site 1 is a substitutional trapping site,
whereas site 3 may be an octahedral interstitial trapping site. Besides quantitatively describing their spectra this model qualitatively explains also the observed annealing behavior and the preparation dependence of the three sites.

As in the argon matrices site 2 could be observed only in those krypton samples
prepared by a "high energy" technique, a fact that holds true also for xenon crys-

tals. Foner et al. /1960/ showed in their analysis that the observed lineshape can be explained by the super hyperfine interaction of the hydrogen S electron with the nuclei of the surrounding xenon atoms. They showed that similar to Ar:H for Xe:H site 1 should be substitutional and site 2 an octahedral interstitial trapping site.

Jackel et al. /1968/ proved that ESR experiments on RGMI hydrogen atoms can indeed be described in terms of the above spin Hamiltonian. They showed that the results are identical for hydrogen and deuterium atoms and that their results at 24 GHz agreed with all previous experiments at 9 GHz, confirming that the parameters of the spin Hamiltonian are not frequency dependent.

Alkali metals. After hydrogen, the alkali metals are RGMI impurities that have been the most extensively studied, mainly by Jen et al. /1962/. Due to multiple trapping sites and super hyperfine interaction between impurity and rare gas matrix in several systems, the lines in the ESR spectra were unresolved. Therefore these authors give only minimum and maximum values for the observed hyperfine and g factor shifts and, if the hyperfine structure could be at least partially resolved, also the number of trapping sites.

They observed (see Table 4.2) two trapping sites for Li in argon, as well as in krypton. Due to unresolved super hyperfine interaction caused by the surrounding xenon nuclei for Xe:Li the number of different trapping sites is uncertain. Site 2 for Ar:Li could be confirmed by Goldsborough and Koehler /1964/ for both Li isotopes.

For Ar:Na (see Table 4.3) Goldsborough and Koehler found a total of five well defined trapping sites. Jen and his co-workers saw even six different trapping sites for this system. For Kr:Na these authors observed only two, but well resolved sites whereas in the xenon matrix the structure of the spectrum was unresolved.

Table 4.2. Hyperfine and g factor shifts for RGMI lithium atoms on various trapping sites

Matrix	Site	$\frac{\Delta A}{A_0} \cdot 10^2$	$-\Delta g \cdot 10^3$	Sample Preparation	Author
Ar	1	-1.6	0.5	C	a
	2	3.1	1.3	C	a
		3.3	2.5	C	b
		3.4	3.1	C	b
Kr	1	-1.7	3.6	C	a
	2	2.2	5.7	C	a
Xe	1	-1.2	10.9	C	a

a   Jen et al. /1962/
b   Goldsborough and Koehler /1966/

Table 4.3. Hyperfine and g factor shifts for RGMI sodium atoms on various trapping sites

Matrix	Site	$\frac{\Delta A}{A_0} \cdot 10^2$	$-\Delta g \cdot 10^3$	Sample Preparation	Author
Ar	1	-2.15	2.6	C	a
	2	-1.70	3.1	C	a
	3	-0.83	1.0	C	a
	4	0.76	1.8	C	a
	5	4.82	2.3	C	a
Kr	1	-1.4	4.5	C	b
	2	2.0	9.3	C	b
Xe	1	-1.3	9.8	C	b

a   Goldsborough and Koehler /1964/
b   Jen et al. /1962/

Table 4.4. Hyperfine and g factor shifts for RGMI potassium atoms on various trapping sites

Matrix	Site	$\frac{\Delta A}{A_0} \cdot 10^2$	$-\Delta g \cdot 10^3$	Sample Preparation	Author
Ar	1	-33.5	4.6	C	a
	2	-13.8	3.8	C	a
		-13.4	3.7	N	b
	3	-3.2	3.1	N	b
	4	4.3	2.5	C	a
		5.5	2.5	C	a
		4.7	2.5	N	b
	5	8.9	3.0	C	a
Kr	1	-12.9	10.3	C	a
	2	-1.2	5.9	C	c
	3	6.6	17.4	C	c
	4	17.6	10.3	C	a
Xe	1	-2.5	22.3	C	a
	2	1.7	16.6	C	c

a   Goldsborough and Koehler /1964/
b   Coufal et al. /1974b/
c   Jen et al. /1962/

As for sodium Jen et al. /1962/ and Goldsborough and Koehler /1964/ observed also for potassium (see Table 4.4) in argon matrices a large number of trapping sites.
Six sites are reported by Jen et al. Most of Goldsborough and Koehler's four sites
were observed for both K isotopes, sites 1 and 2 only when the sample was exposed
during deposition to room temperature IR radiation. Their results on multiple trapping sites could be confirmed in part by Coufal et al. /1974b/. These authors observed during annealing experiments that site 2 seems to be stable against annealing,
whereas site 4 disappears after annealing at 25 K and site 3 becomes unstable at 35
K. Coufal and Lüscher /1975/ proposed that site 4 may be a potassium atom on a bcc-
lattice site with 8 neighboring argon atoms, whereas site 3 has 10 nearest argon
neighbors and site 2 was assigned to a fcc-substitutional site with 12 nearest neighbors. With this model they were able to explain the observed spectroscopic and annealing data. For potassium atoms in krypton matrices Jen et al. report two trapping
sites; also Goldsborough and Koehler observed two but different trapping sites. Similarly, for Xe:K both groups report one site respectively. The site reported by Jen
et al. for Xe:K contains an unresolved multiplet.

According to Jen et al., rubidium atoms (see Table 4.5) are trapped in argon matrices on seven different sites and on five different sites in krypton matrices whereas xenon matrices exhibit an unresolved trapping site. In an extensive study on Ar:Rb
Kupfermann and Pipkin /1968/ found, besides the trapping site observed by Goldsborough
and Koehler /1964/, three more sites where at least one of the sites (4) shows different configurations.

Table 4.5. Hyperfine and g factor shifts for RGMI rubidium atoms on various trapping sites

Matrix	Site	$\frac{\Delta A}{A_0} \cdot 10^2$	$-\Delta g \cdot 10^3$	Sample Preparation	Author
Ar	1	-9.5	5.0	C	a
	2	-0.5	1.3	C	b
	3	5.9	3.4	C	b
	4	7.4	4.3	C	b
		7.7	2.4	C	b
		6.1	2.4	C	b
Xe	1	-1.6	2.0	C	c

a Goldsborough and Koehler /1964/
b Kupfermann and Pipkin /1968/
c Jen et al. /1962/

Table 4.6. Hyperfine and g factor shifts for cesium atoms in rare gas matrices

Matrix	Site	$\frac{\Delta A}{A_0} \cdot 10^2$	$-\Delta g \cdot 10^3$	Sample Preparation	Author
Ar	1	0.5	-2.5	C	a
Kr	1	-0.9	1.1	C	a

a   Jen et al. /1962/

Only Jen et al. /1962/ were successful with experiments on RGMI Cs atoms (Table 4.6). They report unresolved spectra for Ar:Cs as well as for Kr:Cs.

Due to the excessive number of trapping sites for most of the alkali atoms and the lack of reliable information on the symmetry and type of these trapping sites as mentioned as inherent disadvantage of ESR experiments on RGMI alkali atoms it would be rather arbitrary to summarize the results by claiming that there are certain correlations between matrix shifts and atomic numbers of matrix materials or impurities. Annealing experiments on Ar:H and Ar:K show, however, clearly that the sites with the largest positive hyperfine shifts assigned to interstitial trapping sites anneal at the lowest observed annealing temperatures whereas the sites with the largest negative hyperfine shifts are most stable against annealing and are assigned in both systems to substitutional sites.

Due to the magnetic moment of some abundant xenon isotopes ESR spectra in this matrix material are broadened, due to super hyperfine interaction, often beyond recognition. If the spectra are however, resolved or the line shape can be analyzed, ESR experiments on impurities in xenon matrices give directly the number of xenon atoms neighboring the impurity. This holds true to a lesser degree also for krypton matrices, whereas argon and neon are non-magnetic matrix materials. The lowest reported annealing step for argon is at 12 K /Cochran et al., 1959/. In contrast, neon matrices seem to be unsuitable for the isolation of single impurities for at temperatures normally employed for MIS diffusion in neon is still considerable.

Cu, Ag and Au atoms. Isolating metals with a relatively high melting point in a rare gas matrix is a difficult ESR experiment for it employs a high temperature oven for the evaporation of the metal close to the low temperature substrate for the sample deposition. That is why up to now only Kasai and McLeod /1971/ reported results on these systems (Tables 4.7,8,9).

Copper atoms in neon matrices are trapped as easily in amorphous regions of the sample (site 5) as on well defined trapping sites. The four sites were observed for both Cu isotopes and showed a unique dependence of the sample orientation, demon-

Table 4.7. Hyperfine and g factor shifts for RGMI copper atoms on various trapping sites

Matrix	Site	$\frac{\Delta A}{A_0} \cdot 10^2$	$-\Delta g \cdot 10^3$	Sample Preparation	Author
Ne	1	-32.0	7.6	C	a
	2	-31.0	7.6	C	a
	3	-22.6	11.4	C	a
	4	-21.8	11.4	C	a
	5	2.0	3.0	C	a
Ar	1	4.8	2.9	C	a
Kr	1	3.0	6.8	C	a
Xe	1	0.5	8.1	C	a

a   Kasai and McLeod /1971/

Table 4.8. Hyperfine and g factor shifts for silver atoms in rare gas matrices

Matrix	Site	$\frac{\Delta A}{A_0} \cdot 10^2$	$-\Delta g \cdot 10^3$	Sample Preparation	Author
Ne	1	1.3	0.0	C	a
Ar	1	5.7	2.5	C	a
Kr	1	3.8	8.1	C	a
Xe	1	1.0	10.1	C	a

a   Kasai and McLeod /1971/

Table 4.9. Hyperfine and g factor shifts for RGMI gold atoms on various trapping sites

Matrix	Site	$\frac{\Delta A}{A_0} \cdot 10^2$	$-\Delta g \cdot 10^3$	Sample Preparation	Author
Ne	1	1.2	0.7	C	a
	2	1.3	0.6	C	a
Ar	1	2.8	1.1	C	a
Kr	1	1.4	6.1	C	a
Xe	1	-0.9	5.3	C	a

a   Kasai and McLeod /1971/

strating that the direction of sample growth in this system is an axis of lower symmetry for the impurity. In their analysis Kasai and McLeod show that pair one and
two as well as the pair of sites 3 and 4 may be caused by copper atoms on sites with
different numbers of nearest neighbors or different crystal modifications, where the
small difference within each pair is attributed to different configurations of these
nearest neighbors. For copper atoms in argon, krypton and xenon matrices they observe
no multiple trapping sites. The signals are sharp and isotropic. In the case of Xe:Cu
they show that due to the line shape Cu must have 12 nearby Xe nuclei and therefore
the trapping site is a substitutional site. All these effects are observed for both
copper isotopes.

In all rare gas matrices, silver atoms seem to be trapped on only one site. The
lineshape for Kr:Ag and Xe:Ag caused by super hyperfine interaction with the surrounding magnetic matrix nuclei indicates that also in these cases the impurity is trapped
on a substitutional trapping site; a result that was obtained for both Ag isotopes.

Similarly, Ne:Au shows one pair of slightly different trapping sites, whereas in
all other matrices no multiple trapping sites were found for gold atoms. Also in this
case a careful analysis of the line shape indicates that at least for Kr and Xe gold
is trapped on an undistorted substitutional lattice site.

These experiments demonstrate clearly that for each of the impurities the hyperfine shift decreases whereas the g factor shifts are progressively larger in the order of Ar, Kr and Xe. On the other hand, Ag shows the largest shift when comparing
Cu, Ag and Au for one matrix gas. Neon does not fit in these series due to the multiple trapping on sites that are not substitutional.

4.1.2 Atoms with $(np)^3$ Electron Configuration

As for group IA and IB metals extensive ESR studies were also performed on RGMI group
VA atoms. These atoms exhibit a $^4S_{3/2}$ ground state symmetry. That is why despite their
different electron configurations group V impurities share the above mentioned disadvantages of all other S ground state elements, i.e., no reliable data on the symmetry
of the trapping sites is available.

ESR spectra of RGMI nitrogen atoms (Table 4.10) were first reported by Wall et al.
/1958,1959/. They irradiated N_2 doped rare gas polycrystals with γ radiation from a
^{60}Co source. Their results show that N is trapped on multiple trapping sites. The
number of different types of sites as well as the distribution of N atoms over these
sites is reported to depend largely on details of the sample deposition. Adrian et
al. /1962/ produced Ar:N by UV photolysis of Ar:NO; numerical data on the parameters
of the spin Hamiltonian could not be derived from the published spectrum. Similarly
Fischer et al. /1968/ doped various types of matrices with N via the photolytic decomposition of NH_3. Their results indicate a rapidly increasing matrix shift pro-

Table 4.10. Hyperfine and g factor shifts for RGMI nitrogen atoms on various trapping sites

Matrix	Site	$\frac{\Delta A}{A_0} \cdot 10^2$	$-\Delta g \cdot 10^3$	Sample Preparation	Author
Ar		~10.0		γ	a
		13.7	0.9	P	b
	1	18.0	0.1	γ	c
Kr	2	19.8	1.0	P	b
Xe	1	~10.0		γ	a
	2	18.8	0.3	γ	c
	3	31.4		P	b

a Wall et al. /1959/
b Fischer et al. /1967/
c Jackel et al. /1968/

Table 4.11. Hyperfine and g factor shifts for RGMI phosphorous atoms on various trapping sites

Matrix	Site	$\frac{\Delta A}{A_0} \cdot 10^2$	$-\Delta g \cdot 10^3$	Sample Preparation	Author
Ar	1	46.6		P	a
		46.3	0.7	γ	b
	1	51.5	2.0	γ	b
Kr	2	55.2	-0.6	γ	c
Xe	1	57.3		γ	b

a Adrian et al. /1962/
b Jackel et al. /1968/
c Morehouse et al. /1966/

ceeding from Ar to Kr and Xe. Jackel et al. /1968/ irradiated rare gas crystals containing NH_3 with γ radiation to produce RGMI nitrogen atoms. Their result seems to be contradictory. They observed no major change in the matrix shift when varying the matrix material.

Adrian et al. /1962/ report the first ESR spectrum of RGMI phosphorous atoms (Table 4.11). They produced their samples by UV photolysis of PH_3 doped argon crystals. The hyperfine shift could be derived from the published spectra. The results

agree very well with later data from Jackel et al. /1968/ on Ar:P. Jackel and co-
workers verified also results by Morehouse et al. /1966/ on Kr:P and conducted experiments on Xe:P. Jackel et al. used the γ induced decomposition of RGMI PH_3 to
dope their samples with atomic phosphorous. Their results indicate that the hyperfine shift of phosphorous atoms increases with increasing atomic number of the matrix material.

For arsenic atoms (Table 4.12) Jackel et al. /1968/ as well as Morehouse et al.
/1966/ observed large negative hyperfine shifts in rare gas matrices; all samples
had been prepared by γ irradiation of AsH_3 doped crystals. A certain relationship
between matrix shift and matrix material as in the previous discussed case of RGMI
phosphorous atoms cannot be claimed.

Table 4.12. Hyperfine and g factor shifts for arsenic atoms in rare gas matrices

Matrix	Site	$\frac{\Delta A}{A_0} \cdot 10^2$	$-\Delta g \cdot 10^3$	Sample Preparation	Author
Ar	1	-47.2	0.5	γ	a
Kr	1	-52.7	-1.0	γ	b
		-52.7	1.4	γ	a
Xe	1		2.2	γ	a

a   Jackel et al. /1968/
b   Morehouse et al. /1966/

4.1.3 Atoms and Ions with $(3d)^5(4s)^n$ Electron Configuration

In a series of experiments on rare gas matrices containing both electron donating
and electron accepting species Kasai /1968/ showed that charged species created by
UV irradiation of these samples can be isolated and effectively trapped in a rare
gas matrix.

Using Cd, Cr, Mn and Na atoms as electron donors and HI as acceptor he was successful in observing the ESR spectra of the isolated metal atoms of Cr and Mn before the UV irradiation and the RGMI ions Cd^+, Cr^+ and Mn^+ after irradiation. No
reliable free atomic or ionic values are available for these species; therefore
Kasai's original experimental results are summarized in Table 4.13.

Table 4.13. Hyperfine splitting constant and g factor of impurities with $(3d)^5(4s)^n$ electron configuration trapped in argon matrices

Impurity	$A_{eff}/10^6 \ s^{-1}$	g_{eff}	Sample Preparation	Author
$^{111}Cd^+$	14385	2.0006	P	a
$^{113}Cd^+$	15048	2.0006	P	a
^{53}Cr	94.6	2.0006	C	a
$^{53}Cr^+$	67.2	2.0011	P	a
^{55}Mn	78.2	2.0013	C	a
$^{55}Mn^+$	757.9	2.0024	P	a

a   Kasai /1968/

4. 2 Paramagnetic Impurities with P Ground State Symmetry

ESR experiments on RGMI Group III metal atoms sharing the common property of an s^2p^1 electron configuration are of particular interest. Due to their unpaired p electron outside closed shells these atoms are extremely reactive and form strong van der Waals complexes with rare gas atoms. Large van der Waals binding energies and substantial axial splittings would be expected for these atoms in rare gas matrices.

In an extensive study Ammeter and Schlosnagle /1973b/ report results on Al and Ga atoms in various rare gas matrices (Tables 4.14,15). As in the Ar:Cu experiments of Kasai and McLeod /1971/, Knight and Weltner /1971/ using RGMI Al and Ammeter and Schlosnagle /1973/ using RGMI Ga atoms also observed a strong dependence of the ESR spectra on the sample orientation (Fig.4.2.). This indicates that the species is highly oriented in the matrix relative to the direction of sample growth, thus, allowing the determination of both components $A_{eff\perp}$ and $A_{eff\parallel}$ of the hyperfine tensor and of $g_{eff\perp}$ and $g_{eff\parallel}$ of the corresponding g_{eff} tensor.

Knight and Weltner assigned their results originally to an Al-X complex where X was assumed to be an additional type of impurity; an Al_3O rare gas complex had been proposed. But Ammeter and Schlosnagle showed that an interpretation on the basis of different trapping sites for Al atoms is more realistic. Following the interpretation of Ammeter and Schlosnagle aluminum atoms are trapped in neon as well as in argon matrices on two trapping sites that are only slightly different but all highly anisotropic. For Kr:Al they report in addition to sites 1 and 2 (the difference between these sites as well as their anisotropy is smaller than in the neon matrix) an additional site 3 which shows larger matrix shifts. In the xenon matrix, due to line broadening by super hyperfine interaction with the matrix only one trapping

Table 4.14. Components of the hyperfine and g tensors for RGMI aluminum atoms on various trapping sites

Matrix	Site	Direction	$A_{eff}/10^6$ s^{-1}	g_{eff}	Sample Preparation	Author
Ne	1	‖	138.9	2.000	C	a
		⊥	-105.9	1.925	C	a
	2	‖	138.9	2.000	C	a
		⊥	-105.9	1.927	C	a
Ar	1	‖	143.1	2.000	C	a
		⊥	102.0	1.951	C	a
	2	‖	143.1	2.000	C	a
		⊥	101.1	1.956	C	a
Kr	1	‖	135.9	2.001	C	a
		⊥	90.0	1.989	C	a
	2	‖	101.1	2.001	C	a
	3	‖	174.0	1.997	C	a
		⊥	75.9	1.962	C	a
Xe		‖	140.1	2.001	C	a
		⊥	Δ75	2.020	C	a

a   Knight and Weltner /1971/ and
    Ammeter and Schlosnagle /1973b/

Table 4.15. Components of the hyperfine and g tensors for RGMI gallium atoms on various trapping sites

Matrix	Site	Direction	$A_{eff}/10^6$ s^{-1}	g_{eff}	Sample Preparation	Author
Ar	1	‖	434.91	1.9396	C	a
		⊥	-601.86	1.5805	C	a
	2	‖	432.39	1.9456	C	a
		⊥	-594.06	1.6015	C	a
Kr	1	‖	395.55	1.9522	C	a
		⊥	-542.37	1.6751	C	a
	2	‖	392.58	1.9602	C	a
		⊥	-531.09	1.7014	C	a
Xe		‖	342.00	1.968	C	a

a   Ammeter and Schlosnagle /1973b/

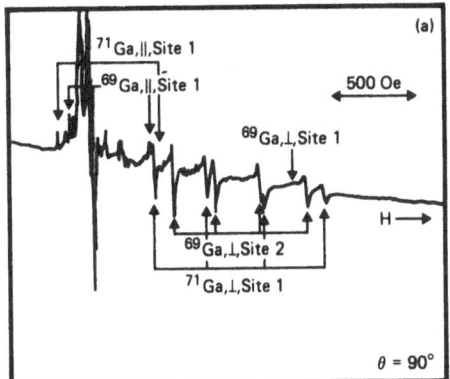

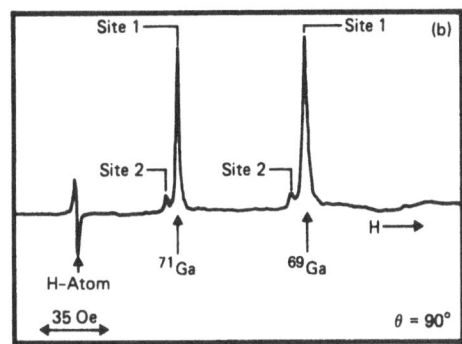

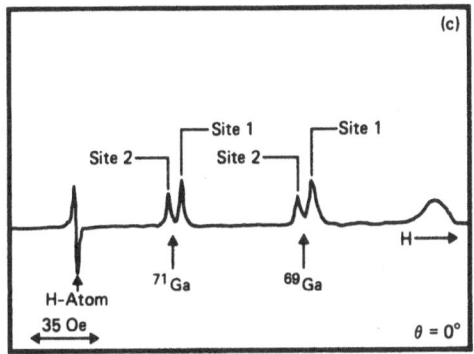

Fig.4.2. The ESR spectra of gallium atoms isolated in argon at 4.2 K after annealing showed strong orientational dependence: (a) full spectrum at θ = 90°, (b) parallel lines at the low field end of spectrum at θ = 90°, (c) same field range as spectrum (b) but at an angle θ = 0° /Ammeter and Schlosnagle, 1973b/

site could be assigned. For aluminum atoms in all matrices and on all trapping sites $|A_{\parallel}| \gg |A_{\perp}|$ and $g_{\parallel} > g_{\perp}$ was observed.

For gallium atoms in argon and krypton matrices Ammeter and Schlosnagle also report multiple trapping sites. Only two slightly different sites are observed for each matrix. Here $g_{\parallel} \gg g_{\perp}$ holds true but $|A_{\parallel}| \ll |A_{\perp}|$ is found. Only an approximate value for one trapping site of Xe:Ga is available due to unresolved spectra.

Ammeter and Schlosnagle also determined the temperature shift of the magnetic parameters; a shift that might be due to annealing. For Ar:Al these shifts seem to be negligible. For Ar:Ga as well as Kr:Ga however the perpendicular lines of the spectrum are strongly temperature dependent. Changes of several percent for $A_{eff\perp}$ as well as for $g_{eff\perp}$ were observed while the parallel lines in the spectra remained almost unchanged. In all cases annealing of the matrix above 25-30 K for Ar and 40 K for Kr caused a rapid loss in the intensity of the ESR signal due to the onset of diffusion. In their elaborate theoretical interpretation that will be discussed in more detail below Ammeter and Schlosnagle show that Al and Ga atoms are trapped in rare gas matrices on axially distorted substitutional lattices sites.

Table 4.16. Components of the hyperfine and g tensor for iodine atoms trapped in xenon matrices

Direction	$A_{eff}/10^6$ s^{-1}	g_{eff}	Sample Preparation	Author
‖	889	1.400	P	a
⊥	1605	2.532	P	a

a   Iwasaki et al. /1979/

Iwasaki et al. /1979/ reported the first clear ESR experiment on RGMI halogen atoms; halogen atoms with a (np)^5 electron configuration show $^2P_{3/2}$ ground state symmetry. Samples were prepared by the UV photolysis of Xe:HI at 4.2 K. g-, hyperfine- and quadrupolar-coupling tensors (Table 4.16) are reported to be axially symmetric and consistent with atomic beam data. The features of the spin Hamiltonian are interpreted in terms of a simple crystal field model.

Furthermore, the effect of the matrix wave functions and the dynamic Jahn-Teller effect are taken into consideration. Iwasaki et al. suggest that earlier results by Ogren and Willard /1971/ on the products of a self-radiolysis of tritiated ethyl iodine C_2H_4TI trapped in xenon matrices can be partially assigned to RGMI iodine.

4.3 Double Resonance Experiments

ESR experiments on RGMI atoms with S ground state symmetry proved to be very powerful in the identification of a species by its typical hyperfine pattern and in assigning different trapping sites to different hyperfine multiplets. But no convincing data on the symmetry of these sites can be derived from ESR spectra. Optical absorption spectra of the same RGMI species show a complicated structure in part due to multiple trapping and in part due to lifting the degeneracy of the excited states by interaction with the matrix. Combining both spectroscopic tools, ESR that gives information on the ground state of the impurity and optical absorption spectroscopy that deals with transmissions from this ground state to excited states, should help to overcome the shortcomings of each technique. The idea was thus to assign trapping sites, as identified by ESR, in some double resonance experiment to the optical absorption bands. In a second step the symmetry at these sites, as derived from optical splittings, can be determined. But in realizing such experiments severe difficulties are encountered. The first problem is to produce a sample that is suitable for ESR as well as for optics and the second problem is to fit an optical setup into an ESR magnet.

That is why up to now only two partially successful experiments have been reported. Kupfermann and Pipkin /1968/ performed a series of optical absorption and ESR experiments on Ar:Rb. They monitored the difference in the absorption of left and right circularly polarized light. This difference is proportional to the difference in the population of the Zeeman levels of the 5s ground state levels of the Rb atoms. Saturating the ESR transition between these Zeeman levels would equalize the population of these ground state levels thus changing the optical signal. The magnetic field was swept while irradiating with constant microwave frequency and simultaneously monitoring the difference in absorption between left and right circularly polarized light whose wavelength corresponded to an optical absorption maximum. The resulting double resonance spectrum is shown in Fig.4.3. When the magnetic field hits the ESR line of those atoms that cause the monitored optical absorption, the observed double resonance signal should decrease.

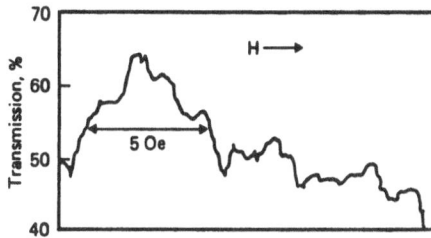

Fig.4.3. Double resonance experiment on Ar:Rb. The chart recording shows the difference in the absorption of left and right circularly polarized light of a wavelength of 713 nm versus magnetic field while irradiating at constant microwave frequency /Kupfermann and Pipkin, 1968/

With this experiment Kupfermann and Pipkin /1968/ were able to assign trapping site 4 of Ar:Rb to the longest wavelength component of the blue-shifted optical absorption triplet. Furthermore, they proved that this ESR line can be saturated and is homogeneously broadened.

In a similar double resonance experiment on Ar:K Coufal and Lüscher /1974/ were successful in assigning two trapping sites to optical absorption bands. The intensity of an ESR line on RGMI potassium atoms is proportional to the number of atoms on that particular type of trapping site that are in their 4s ground state. By irradiating into the optical absorption band of these atoms the population in that 4s ground state can be slightly reduced, thus reducing the amplitude of that ESR line. In a first type of experiment Coufal and Lüscher monitored a particular ESR line while irradiating with chopped high intensity light into a small band within the optical absorption spectrum. The relative modulation amplitude of the ESR lines due to the light modulation at the indicated wavelength is shown in Fig.4.4. They conclude

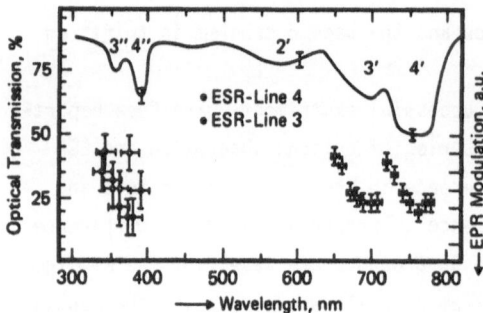

Fig.4.4. Double resonance experiment on Ar:^{41}K. ESR correlation signal for ESR lines corresponding trapping site 3(o) and 4(•) when irradiating with light of the indicated wavelength. In addition the optical transmission spectrum of the same sample is shown /Coufal and Lüscher, 1974; Coufal et al., 1974a/

that the optical absorption bands 4' and 3' are caused by potassium atoms with different ground states, i.e., on different trapping sites. This figure emphasizes that bands 4' and 4" respectively band 3' and 3" belong to the 4p←4s and 5p←4s transitions of potassium atoms on two types of trapping sites. This experiment gave also a rough estimate for the spin-lattice relaxation time T_1

$$10^{-7} s \leq T_1 \leq 10^{-3} s \quad .$$

Less meaningful but eventually more convincing was an additional experiment by Coufal and Lüscher. They used chopped broad-band light irradiation into one optical absorption band to modulate the ground state population of this particular absorption band. Scanning a normal ESR spectrum only that ESR line that corresponds to that particular optical absorption band should be observed, but as a second derivative instead of the normally noted first derivative. The results of this experiment are shown in Fig.4.5.

With these double resonance experiments and by comparing the annealing behavior of ESR line intensities and optical density hyperfine and optical matrix shifts as

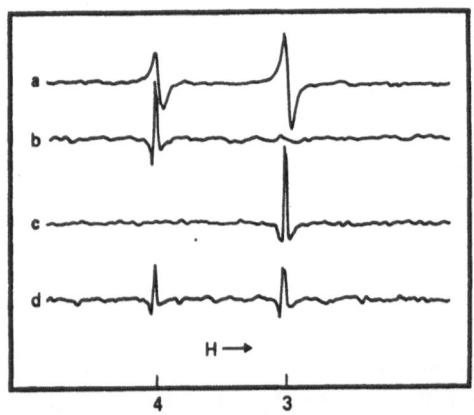

Fig.4.5. Double resonance experiment on Ar:^{41}K. The low field ESR lines of ^{41}K atoms trapped in an argon matrix are shown: (a) without optical irradiation in the first derivative mode and (b) irradiating in optical absorption band 4', (c) irradiating in optical absorption band 3'. (d) irradiating in bands 3" and 4". (b), (c) and (d) are recorded in the second derivative model /Coufal and Lüscher, 1974/

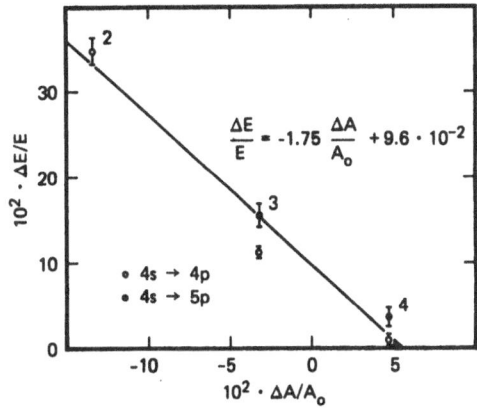

Fig.4.6. Optical matrix shift $\Delta E/E$ versus ESR hyperfine shift $\Delta A/A_0$ for three different trapping sites for potassium atoms in argon crystals /Coufal et al., 1976/

shown in Fig.4.6 could be correlated. The trend for Ar:K seems however contradictory to the observation of Kupfermann and Pipkin for Ar:Rb.

Double resonance experiments thus proved helpful in the interpretation of optical spectra, but the feedback from optical experiments in clarifying the symmetry of multiple trapping sites of RGMI atoms and ions with S ground state symmetry is still missing.

4.4 Models for the Interpretation of the ESR Spectra of RGMI Atoms and Ions

Most experiments have been conducted on RGMI impurities with S ground state symmetries whereas only few have been performed on impurities with P ground states. That is why models for the first class of impurities are widely reported in the literature whereas only one model specific for RGMI P ground state impurities has been put forward. An atom or ion trapped in a rare gas matrix undergoes various interactions with the surrounding matrix. Perturbations of the Zeeman and hyperfine energies are a consequence. In order to calculate these perturbations details of the physical configuration of the trapping sites would be necessary.

4.4.1 Models for Impurities with S Ground State Symmetry

For RGMI impurities with S ground state symmetry no details of the configuration of the various trapping sites can directly be derived from the ESR spectra. That is why a relatively simple model by Adrian /1960/ became popular in this field. This model is based on the following assumptions:

1) The contributions of the matrix atoms to the matrix shift of the impurity are additive. The many body problem of calculating these shifts is reduced to a pairwise interaction. The number of nearest neighbors to the impurity is a parameter that can be used in the interpretation of spectra.

2) The interaction between the impurity and one matrix atom is assumed to be isotropic. That means that the interatomic distance within this pair is the vari-
   able parameter in this model.

3) Short and long range interactions cause additive contributions to the matrix
   shifts and can therefore be accounted for separately.

These assumptions are identical to those used to justify the two body, pairwise additive interatomic potential to describe the properties of pure rare gas crystals.
As in this application also the long range contribution to the matrix shift is due
to the van der Waals interaction whereas short range effects are caused by overlap
(Pauli exclusion) and perturbation effects due to large distortions of the electronic
wave functions. In a perturbation theory approach the final results for the matrix
shifts can be written in the form

$$\Delta g = \frac{8}{9} \frac{\lambda_{p\sigma}}{E_i} \, |<\psi_i|\psi_{p\sigma}>|^2 \left[1 + \frac{2}{E_i} f(r)\right] \tag{4.6}$$

and

$$\frac{\Delta A}{A_0} = -\frac{2}{E_i} \cdot V(r) + \frac{1}{E_i + E_M} \cdot g(r) + \sum_m |<\psi_i|\psi_{Mm}>|^2 \tag{4.7}$$

with $V(r)$ the potential between impurity and one matrix atom, consisting of a repulsive term $f(r)$ and an attractive part $g(r)$

$$V(r) = f(r) + g(r) \quad .$$

E_i and E_M are mean excitation energies of impurity and matrix atom. $\lambda_{p\sigma}$ stands for
the spin orbit splitting constant of the outermost $p\sigma$ orbital $\psi_{p\sigma}$ of the matrix
atom.

Figure 4.7 shows a graphical representation of these results for Ar:H. The g-factor shift, caused by overlap and the repulsive part of the potential, becomes larger
at smaller interatomic distances. The hyperfine shift $\Delta A/A_0$ is dominated by the interatomic potential and its dependence on the interatomic distance. At very short
distances overlap and interatomic potential "squeeze" the unpaired electron in the
nucleus of the impurity thus increasing the hyperfine coupling. At longer interatomic distances attractive van der Waals forces tend to drag the electron away
therefore decreasing the spin density at the impurity nucleus.

For comparison with experimental results or to assign trapping sites to various
ESR multiplets for all conceivable configurations, i.e., number of rare gas atoms
surrounding an impurity and their distances from the impurity hyperfine and g-factor shifts are calculated by adding the contributions of the various neighbors.

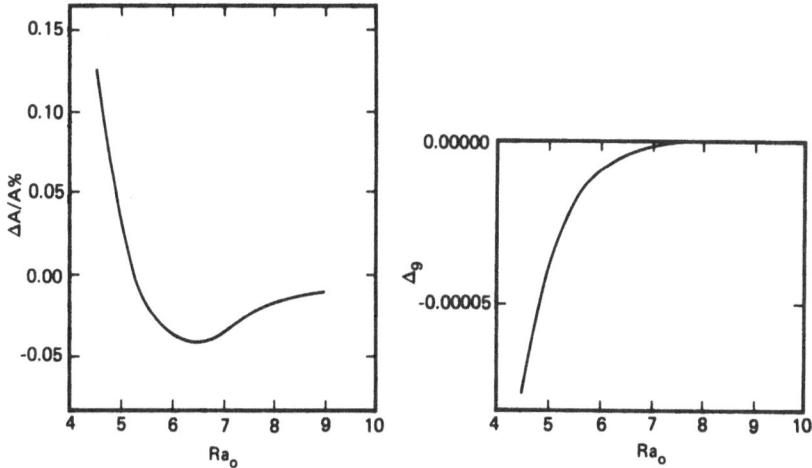

Fig.4.7. The effect of one argon atom on hyperfine and g-factor shift of a trapped hydrogen atom /Adrian et al., 1962/

This type of model is not at all sensitive to the symmetry of the trapping site and allows mainly a determination of the density of the rare gas cage but it should be clearly possible to distinguish between configurations like substitutional versus interstitial trapping sites.

Originally, Adrian proposed his semiempirical model in 1960 for the qualitative interpretation of the ESR spectra of hydrogen atoms in matrices. He extended his calculations on Ne:N and Ar:N /Adrian, 1962/. A slightly improved version was used by Jen et al. /1962/ and Coufal and Lüscher /1975/ for the interpretation of their spectra of RGMI alkali atoms. Jackel et al. /1968/ further modified the basic model to discuss their experimental results on N and As in rare gas matrices. For comparison with his own model Smith /1964/ recalculated the matrix shifts for hydrogen and alkali atoms in various rare gas matrices on the basis of Adrian's original model and a modified model; both calculations use more recent values for wave functions and other parameters.

In his variational theory of paramagnetic impurities in van der Waals crystals Smith /1963/ advanced the theory of RGMI impurities with S ground state symmetry. He derived the ground state wave function and the ESR parameters of a tight binding paramagnetic center in a rare gas crystal from first principles. Starting from a one electron, tight binding static lattice picture of the impurity doped crystal variational corrections accounting for perturbations due to the crystal field, spin-orbit interaction and van der Waals interaction are added. His results show that the g-factor shift arises almost exclusively from the spin-orbit interaction correction to the rare gas wave function and to overlap of impurity rare gas matrix orbitals. The hyperfine shift arises from renormalization associated with orthogonalization of the one electron wave functions and both crystal field and van der Waals corrections. Despite the quite different approach, his final results are in excellent agreement with Adrian's perturbation theory. Smith's /1964/ numerical evaluation of his theory for the g-factor shifts of RGMI hydrogen and alkali metals are in close agreement with experimental results as are his calculations based on Adrian's original as well as on a modified model.

4.4.2 Models for Impurities with P Ground State Symmetry

Besides their excellent experimental work on Al and Ga atoms in rare gas matrices in determining the parameters of the g_{eff} and A_{eff} tensors of the trapped species, Ammeter and Schlosnagle /1973/ also advanced the theoretical understanding of RGMI impurities.

Starting from a static crystal field model they assigned the observed ESR spectra of Al and Ga unambiguously to impurity atoms on axially symmetric sites that all but site 3 for Kr:Al are proven to be centrosymmetric. A static MO approach was not in full agreement with their experimental results. Their final non-adiabatic MO model however proved to be even quantitatively consistent with all observed effects.

Their conclusions are that gallium as well as aluminum atoms are trapped on substitutional sites within a rare gas lattice and that these sites undergo a static distortion along one of the three equivalent tetragonal axis due to Jahn-Teller coupling. Occupation of a substitutional site in a hcp surrounding distorted along its hexagonal axis could not be ruled out and may be an explanation for the observed pairs of trapping sites.

4.5 Summary

ESR spectroscopy of RGMI atoms and ions is successful in identifying trapped species by their characteristic hyperfine pattern and in distinguishing different trapping sites. Double resonance experiments correlate these results with those obtained by other spectroscopic techniques.

The interpretation of ESR spectra shows that in many cases of impurity trapping in rare gas matrices substitutional sites are occupied, furthermore, octahedral interstitial sites seem to be present when comparatively small atoms like, for example, hydrogen or lithium are trapped. The more complicated trapping sites observed for some systems are probably sites with more than twelve nearest rare gas neighbors or sites with impurity-vacancy or as demonstrated by Hauge et al. /1978/, for $Ar:Li:H_2O$ impurity-impurity complexes. In all these cases ESR spectra largely depend on sample preparation and handling techniques.

References

Adrian, F.J. /1960/: J. Chem. Phys. *32*, 972 (Sect.4.4.1)
Adrian, F.J. /1962/: Phys. Rev. *127*, 327 (Sects.4.1.2, 4.4.1)
Adrian, F.J., E.L. Cochran, V.A. Bowers /1962/: Advn. Chem. Soc. *36*, 50 (Sect.4.1.2)
Ammeter, J.H., D.C. Schlosnagle /1973a/: Chimia *27*, 372 (Sects.4, 4.4.2)
Ammeter, J.H., D.C. Schlosnagle /1973b/: J. Chem. Phys. *59*, 4784 (Sects.4, 4.1.3, 4.4.2)
Ayscough, P.B. /1967/: *Electron Spin Resonance in Chemistry* (Methuen, London) p.438 (Sect.4)

Baker, J.M., B. Bleaney, W. Hayes /1958/: Proc. Roy. Soc. A*247*, 141 (Sects.3.1.3, 3.3.5)
Batchelder, D.N., M.F. Collins, B.C. Haywood, G.F. Sidey /1976/: J. Phys. C *3*, 249 (Sect.3.3.3)
Belyaeva, A.A., Y.B. Predtechenskii, L.D. Sherba /1969/: Izv. Akad. Navk. SSSR Ser. Fiz. *33*, 895 (Sect.3.3.5)
Belyaeva, A.A., Y.B. Predtechenskii, L.D. Sherba /1973/: Opt. Spectrosc. *34*, 21 (Sect.3.3.5)
Blaise, J., A. Steudel /1968/: Z. Physik *209*, 311 (Sect.3.3.1)
Bouldin, W.V., W. Gordy /1964/: Phys. Rev. A *135*, 806 (Sects.1, 2.1.2, 2.3.1, 4.1.1)
Bos, A., A.T. Howe /1974/: J. Chem. Soc. Faraday Trans. *70*, 451 (Sect.3.3.1)
Brix, P., S. Hüfner, P. Kienle, D. Quitmann /1964/: Phys. Lett. *13*, 140 (Sect.3.3.1)

Childs, W.Y., L.S. Goodman /1966/: Phys. Rev. *48*, 74 (Sects.3.1.2, 3.3.2)
Cochran, E.L., V.A. Bowers, S.N. Foner, C.U. Jen /1959/: Phys. Rev. Lett. *2*, 43 (Sect.4.1.1)
Coulthard, M.A. /1973/: J. Phys. *36*, 23 (Sect.3.3.1)
Coufal, H., U. Nagel, M. Burger, E. Lüscher /1974a/: Phys. Lett. A*47*, 327 (Sects. 2.3.1,2)
Coufal, H. M. Burger, U. Nagel, E. Lüscher, K. Böning, G. Vogl /1974b/: Phys. Lett. A*48*, 143 (Sects.2.1.2, 2.3.1,2, 4.1.1)
Coufal, H., E. Lüscher /1974/: Phys. Lett. A*48*, 445 (Sect.4.3)
Coufal, H., E. Lüscher /1975/: Proc. 14. Low Temp. Phys. LT14 *3*, 44; ed. by M. Krusius, M. Vuorio (North Holland) (Sects.4.1.1, 4.4.1)
Coufal, H., U. Nagel, M. Burger, E. Lüscher /1976/: Z. Physik B *25*, 227 (Sect.2.3.1)
Coufal, H., U. Nagel, E. Lüscher /1978/: Ber. Bunsenges. Phys. Chem. *82*, 133 (Sect. 2.1.2)

de Vries, J.L.K.F., J.M. Trooster, P. Ros /1975/: J. Chem. Phys. *63*, 5256 (Sect.3.3.1)
Duerig, W.H., I.L. Mador /1952/: Rev. Sci. Instr. *23*, 421 (Sect.2.3.2)
Duff, K.J. /1974/: Phys. Rev. B *9*, 66 (Sects.3.1, 3.2.1,3)

Fischer, P.H.H., S.W. Charles, C.A. McDowell /1967/: J. Chem. Phys. *46*, 2162 (Sects. 4.1.1,2)
Foner, S.M., E.L. Cochran, V.A. Bowers, J.K. Ken /1960/: J. Chem. Phys. *32*, 963 (Sect.4.1.1)
Fujii, Y., N.A. Lurie, G. Shirane /1974/: Phys. Rev. B *10*, 3637 (Sect.3.3.3)

Gerth, G., P. Kienle, K. Luchner /1968/: Phys. Lett. A *27*, 557 (Sect.3.3.1)
Gerth, G., K. Luchner, H. Micklitz /1972/: Phys. Stat. Sol. (b) *53*, 593 (Sect.2.1.2)
Goldsborough, J.P., T.R. Koehler /1964/: Phys. Rev. *133*, A135 (Sects.2.3.1, 4.1.1)
Griffith, J.S. /1964/: *The Theory of Transition Metal Ions* (Cambridge University Press, Cambridge) p.393 (Sect.3.1.2)
Gruen, D.M. /1976/: *Cryochemistry*, ed. by M. Moskovits, G.A. Ozin (Wiley, New York) p.482 ff (Sect.4)

Handbook of Chemistry and Physics, 59th Edition, 1978 (CRC Press, West Palm Beach) F-252 (Sect.4)
Hauge, R.H., P.F. Meier, J.I. Margrave /1978/: Ber. Bunsenges. Phys. Chem. *82*, 102 (Sect.4.5)

Hüfner, S., H.H. Wickmann, C.F. Wagner /1968/: Hyperfine Structure and Nuclear Radiation, ed. by E. Mathias, D.A. Shirley (North Holland, Amsterdam) p.952 (Sect. 3.3.4)

Iwasaki, M., K. Toriyama, H. Muto /1979/: J. Chem. Phys. *71*, 2853 (Sects.4, 4.2)

Jaccarino, V., G.K. Wertheim /1962/: Proc. 2nd Intern. Conf. Mössbauer Effect, Saclay, France; ed. by D.M.J. Compton, A.H. Schoen (Wiley, New York) p.260 (Sect.1)
Jackel, G.S., W.H. Nelson, W. Gordy /1968/: Phys. Rev. *176*, 453 (Sects.4.1.1,2, 4.4.1)
Jakob, M., H. Micklitz, K. Luchner /1976/: Phys. Lett. A *57*, 67 (Sects.3.1.3, 3.3.3,5)
Jakob, M., H. Micklitz, K. Luchner /1977/: Phys. Lett. A *61*, 265 (Sect.3.1.3)
Jen, C.K., S.M. Foner, E.L. Cochran, V.A. Bowers /1958/: Phys. Rev. *112*, 1169 (Sect.1)
Jen, C.K., V.A. Bowers, E.L. Cochran, S.N. Foner /1962/: Phys. Rev. *126*, 1749 (Sects. 4.1.1, 4.4.1)

Kasai, P.H. /1968/: Phys. Rev. Lett. *21*, 67 (Sects.2.1.2, 4.1.1,3)
Kasai, P.H., D. McLeod Jr. /1971/: J. Chem. Phys. *55*, 1566 (Sects.2.3.1, 4.1.1,3)
Klee, C., T.K.McNab, F.J. Litterst, H. Micklitz /1974/: Z. Physik *270*, 31 (Sect.3.3.1)
Knight, L.B., W. Weltner Jr. /1971/: J. Chem. Phys. *55*, 5066 (Sect.4.1.3)
Kopfermann, H. /1956/: *Kernmomente* (Akad. Verlagsgesellschaft, Frankfurt) p.164 (Sect.3.3.1)
Kupfermann, S.L., F.M. Pipkin /1968/: Phys. Rev. *166*, 207 (Sects.1, 4.1.1, 4.3)
Kusch, P., V.W. Hughes /1959/: "Handbuch der Physik", ed. bv S. Flügge, Vol. 37/I (Springer, Berlin, Heidelberg, New York) pp.100 and 117 (Sect.4)

Lambe, J. /1966/: Phys. Rev. *126*, 1208 (Sect.2.1.2)
Lax, M. /1952/: J. Chem. Phys. *20*, 1752 (Sect.3.3.3)
Litterst, F.J., H. Micklitz, A. Schichl /1976/: Phys. Lett. A *57*, 70 (Sects.3.1.3, 3.3.1,3)
Lurie, N.A., G. Shirane, J. Skalyo /1974/: Phys. Rev. B *9*, 5300 (Sect.3.3.3)

Mann, D.M., H.P. Broida /1971/: J. Chem. Phys. *55*, 84 (Sect.3.1.2)
McNab, T.K., H. Micklitz, P.H. Barrett /1971/: Phys. Rev. B *4*, 3787 (Sects.1, 2.2.2, 3.1.2, 3.3.3,4)
McNab, T.K., P.H. Barrett /1971/: "Mössbauer Effect Methodology". Vol. 7, ed. by I.J. Gruverman (Plenum Publ. Co., New York) pp.59-83 (Sects.1, 3.1.2)
Meyer, B. /1971/: *Low Temperature Spectroscopy* (Elsevier, New York) p.209 (Sect.2.1.1)
Meyer, J., J. Speth /1972/: Phys. Lett. B *39*, 330 (Sect.3.3.1)
Micklitz, H., K. Luchner /1969/: Z. Physik *227*, 301 (Sect.2.1.2)
Micklitz, H., P.H. Barrett /1971/: Phys. Rev. B *4*, 3845 (Sects.3.1.2, 3.3.3)
Micklitz, H., P.H. Barrett /1972a/: Appl. Phys. Lett. *20*, 387 (Sect.3.3.1)
Micklitz, H., P.H. Barrett /1972b/: Phys. Rev. Lett. *28*, 1547 (Sects.2.1.2, 3.2.2, 3.3.3)
Micklitz, H., P.H. Barrett /1972c/: Phys. Rev. B *5*, 1704 (Sects.2.1.2, 3.1.1)
Micklitz, H., F.J. Litterst /1974/: Phys. Rev. Lett. *33*, 1180 (Sects.2.1.2, 3.2.1, 3.3.1)
Micklitz, H., K. Luchner /1974/: Z. Physik *270*, 79 (Sect.3.3.5)
Micklitz, H. /1977/: Hyperfine Interactions *3*, 135 (Sects.3.2.3, 3.3.1)
Micklitz, H. /1981/: "Matrix Isolated Spectroscopy", ed. by A.J. Barnes, W.J. Orville-Thomas, A. Müller, R. Gaufries (Reidel, Dordrecht) p.91 (Sect.1)
Mössbauer, R.L., W.H. Wiedemann /1960/: Z. Physik *159*, 33 (Sect.3.3.3)
Montano, P.A., P.H. Barrett, Z. Shanfield /1974/: Solid State Commun. *15*, 1675 (Sect.3.1.2)
Montano, P.A., P.H. Barrett, Z. Shanfield /1975/: J. Chem. Phys. *64*, 2896 (Sect.3.1.2)
Montano, P.A., P.H. Barrett, H. Micklitz /1976/: "Mössbauer Effect Methodology", Vol. 10, ed. by I.J. Gruverman, C.W. Seidel (Plenum Press, New York) pp.245-260 (Sects.2.1.2, 2.2.1, 3.1.4, 3.3.2)
Montano, P.A., P.H. Barrett, H. Micklitz, A.J. Freeman, J.V. Mallow /1978/: Phys. Rev. B *17*, 6 (Sects.2.1.2, 3.2.1, 3.3.2,3)
Montano, P.A. /1982/: J. Phys. C *15*, 565 (Sect.3.1.3)

Morehouse, R.L., J.J. Christiansen, W. Gordy /1966/: J. Chem. Phys. *45*, 1747 (Sect. 4.1.2)

Mulliken, R.S. /1937/: Phys. Rev. *51*, 310 (Sect.2.1.2)

Moskovits, M., G.A. Ozin (Editors) /1976/: *Cryochemistry* (Wiley, New York) (Sect.1)

Ogren, P.J., J.E. Willard /1971/: J. Phys. Chem. *75*, 3359 (Sect.4.2)

Orbach, R. /1962/: "Fluctuations, Relaxations and Resonance in Magnetic Systems", ed. by D. Haar (Oliver and Boyd, London) p.219 (Sect.3.3.4)

Paul, P., S. Puri /1982/: Phys. Stat. Sol. (b) *114*, 213 (Sect.3.3.3)

Phillips, J.C. /1959/: J. Phys. Chem. Solids *11*, 226 (Sect.3.3.1)

Pleiter, F., B. Kolk /1971/: Phys. Lett. B *34*, 296 (Sect.3.3.1)

Pollack, H. /1962/: Phys. Stat. Sol. *2*, 270 (Sect.2.1.2)

Powell, B.M., G. Dolling /1977/: "Rare Gas Solids", ed. by M.L. Klein, J.A. Venables (Academic Press, London) Vol. II, p.921 (Sect.3.3.3)

Rexroad, H.N., W. Gordy /1962/: Phys. Rev. *125*, 242 (Sect.2.3.2)

Ruby, S.L., G.M. Kalvius, A.B. Beard, R.E. Snyder /1967/: Phys. Rev. *159*, 239 (Sect. 3.3.1)

Ruby, S.L., G.K. Shenoy /1969/: Phys. Rev. *186*, 326 (Sect.3.3.1)

Ruegsegger, R., W. Kündig /1972/: Phys. Lett. B *39*, 296 (Sect.3.3.1)

Ryberg, R., O. Hunderi /1977/: J. Phys. C *10*, 3559 (Sect.2.2.2)

Sanders, P.G.H., G.K. Woodgate /1960/: Proc. Roy. Soc. A *257*, 269 (Sect.3.3.2)

Shamay, S.M., M. Pasternak /1976/: J. Physique Colloq. *37*, C6-525 (Sect.2.2.1)

Sharma, R.R., A.K. Sharma /1972/: Phys. Rev. Lett. *29*, 122 (Sect.3.3.1)

Shenoy, G.K. /1974/: Private communication (Sect.3.3.1)

Simanek, E., A.Y.C Wong /1968/: Phys. Rev. *166*, 348 (Sect.3.3.1)

Skalyo, J, Endoh, J., G. Shirane /1974/: Phys. Rev. B *9*, 1797 (Sect.3.3.3)

Smith, D.Y. /1963/: Phys. Rev. *131*, 2056 (Sect.4.4.1)

Smith, D.Y. /1964/: Phys. Rev. *133*, A1087 (Sect.4.4.1)

Trautwein, A., F.E. Harris, A.J. Freeman, J.P. Deslaux /1975/: Phys. Rev. B *11*, 4101 (Sect.3.3.1)

Van Rossum, M., J. Odeurs, H. Pattyn, J. de Troyer, G. Verbiest, R. Coussement, S. Bukshpan /1980/: Phys. Lett. A *75*, 241 (Sect.3.2.2)

Visscher, W.M. /1960/: Ann. Phys. *9*, 194 (Sect.3.3.3)

Visscher, W.M. /1963/: Phys. Rev. *129*, 28 (Sect.3.3.3)

Walch, P.F., D.E. Ellis /1973/: Phys. Rev. B *7*, 903 (Sect.3.3.1)

Wall, L.A., D.W. Brown, R.E. Florin /1959a/: J. Chem. Phys. *30*, 602 (Sects.1, 2.3.1, 4.1.2)

Wall, L.A., D.W. Brown, R.E. Florin /1959b/: J. Phys. Chem. *63*, 1762 (Sects.1, 2.3.2, 4.1.2)

Wegener, H. /1965/: *Der Mössbauereffect und seine Anwendungen in Physik und Chemie* (Bibliographisches Institut, Mannheim) (Sect.3.3.3)

Nuclear Magnetic Resonance in Condensed Rare Gases

By R.E.Norberg

1. Introduction

Nuclear magnetic resonance has been applied to studies of magnetic nuclides in pure rare gas solids and liquids and of H_2 as a dilute impurity in rare gas hosts. Some of the results in the solids are closely related to those in the liquids and so this review includes the NMR results in both rare gas solids and liquids.

Chapter 2 summarizes NMR results for magnetic nuclides in pure condensed neon, krypton, and xenon. These include diffusion coefficients, quadrupolar spin relaxation, and chemical shifts. Chapter 3 reviews NMR results for dilute H_2 in condensed neon, argon, and krypton. NMR examines the ortho-H_2 component and yields coefficients of interdiffusion, correlation frequencies for molecular transitions, and information about the H_2 sites in the solids.

2. Neon, Krypton, and Xenon

2.1 Spin Relaxation and Diffusion

NMR experiments in condensed neon, krypton, and xenon have been performed on the magnetic nuclides ^{21}Ne, ^{83}Kr, ^{129}Xe, and ^{131}Xe. Some relevant parameters for these species are given in Table 1. The nuclides present an interesting array of properties for NMR experiments. ^{129}Xe has spin $\frac{1}{2}$ and does not experience electric quadrupole interactions. The other species have quadrupole moments and exhibit relaxation times and resonance line broadenings which reflect quadrupole interactions arising from thermal phonons and lattice defects.

Table 1. Magnetic nuclides of neon, krypton, and xenon (based on NMR Table, Varian Associates)

Nuclide	^{21}Ne	^{83}Kr	^{129}Xe	^{131}Xe
Natural Relative Abundance	0.26	11.55	26.24	21.24
Spin ($\hbar$)	3/2	9/2	1/2	3/2
Gyromagnetic Ratio ' $(s\ G)^{-1}$	-2113	-1030	-7402	2193
Electric Quadrupole Moment $(e \cdot 10^{-24}\ cm^2)$	+0.093	+0.15	-	-0.12

2.1.1 Rare Gas Solids

^{129}Xe. Yen /1963/ reported 10.5 MHz pulsed NMR measurements on ^{129}Xe in condensed xenon samples of natural isotopic abundance. The resonance in the cold polycrystalline solid had a Gaussian free induction decay with a spin-spin relaxation time T_2^{RL} of about 1 ms, in agreement with the predicted Van Vleck /1948/ second moment for a fcc lattice, including contributions from both the 129-129 and 129-131 nuclear dipolar interactions. The agreement was confirmed more precisely in the 3 MHz measurements of Warren /1967/ on both free induction decays and solid echoes at 4.2 K.

Above 117 K the ^{129}Xe resonance line is motionally narrowed (Fig.2.1) over three decades of increasing T_2 before the melting point is reached. The T_2 values in the narrowed region were measured /Yen, 1963/ from the envelopes of Carr-Purcell /1954/ echo trains. Throughout the solid temperature range spin-lattice relaxation times

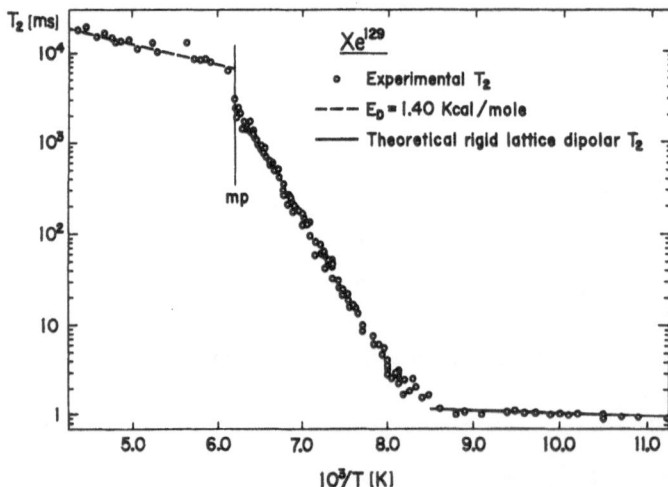

Fig.2.1. Spin-Spin relaxation times for ^{129}Xe in liquid and solid xenon /Yen, 1963/

T_1 in clean xenon samples were very long (greater than 7000 s at 125 K). T_1 was reduced to experimentally more convenient values near 600 s by admitting a small amount of air to the gas samples. The addition did not produce any observed change in the ^{129}Xe $T_2(T)$ results in the solid.

The increase of T_2 with temperature above 117 K corresponds to motional narrowing in the adiabatic regime. The relation between T_2 and τ, the characteristic time for atomic self-diffusion, is thus of the form of the Anderson-Weiss /1953/ expression $1/T_2 \simeq \sigma_D^2 \tau$, where σ_D^2 is the Van Vleck /1948/ dipolar second moment.

In order to deduce precise correlation times and diffusion coefficients from the $T_2(T)$ data it is necessary to take account of the fact that the atomic diffusion takes place on a space lattice, include consideration of correlation effects, and make an assumption about the diffusion mechanism. Wolf /1974a/ has shown that for dipolar interactions and a single nuclear spin species undergoing monovacancy diffusion on a fcc lattice the spin-spin relaxation rate in a powder sample is given by

$$\frac{1}{T_2} = 0.968f \ \alpha\tau(2/a_0)^6 \tag{2.1}$$

which may be rewritten as

$$\frac{1}{T_2} = \frac{4}{3} \sigma_D^2 \tau \quad . \tag{2.2}$$

Here f is the nuclide abundance, a_0 is the fcc cube edge lattice parameter, and $\alpha = \frac{3}{2} \gamma^4 \hbar^2 \ I(I+1)$. Equation (2.2) applies to the ^{129}Xe case using the total σ_D^2 arising from dipolar interactions with both ^{129}Xe and ^{131}Xe and neglecting any isotopic difference in diffusion rates. The results of Yen /1963/, modified via (2.2) by Cowgill /1976/, indicate that self-diffusion in solid xenon between 125 and 161 K follows an Arrhenius relation with

$$D_0 = 9.0 \begin{smallmatrix} +0.5 \\ -0.3 \end{smallmatrix} \ cm^2/s \tag{2.3a}$$

and

$$E_D = 7400 \pm 50 \ cal/mole \quad . \tag{2.3b}$$

There is no evidence of deviation from Arrhenius behavior, even in the vicinity of the xenon triple point.

Chadwick and Morrison /1968,1970/ have reported that the addition of O_2 produced a significant reduction in D_0 for self-diffusion in solid krypton. It is possible that the air added to control the ^{129}Xe T_1 may have affected Yen's xenon diffusion results. However, as stated above, the $T_2(T)$ data did not seem to be changed by the introduction of air. In addition, an NMR search for an O_2 effect on D in solid krypton produced negative results /Cowgill, 1971/. [131]Xe NMR experiments /Warren, 1967/ indicate that the air did not stabilize a hcp xenon phase. The question of the effect of O_2 deserves further study.

Barroilhet /1973,1977/ examined [129]Xe spin relaxation in solid and liquid xenon at 8.2 MHz. Laboratory frame spin-lattice relaxation times T_1 were controlled by an added air impurity and a study was made of $T_{1\rho}$, the spin-lattice relaxation time in the rotating reference frame. A spin-lock was achieved by the method of Slichter and Ailion /1964/. The $T_{1\rho}$ decays were found to exhibit an initial fast component, followed by a slower exponential tail. The initial transient was interpreted as an inhomogeneous rate associated with the random paramagnetic impurities.

In solid xenon above 125 K a long-lived relaxation component was interpreted as a homogeneous dipolar $T_{1\rho}$ associated with the thermally activated atomic diffusion. In the weak collision short τ limit the temperature variation of these $T_{1\rho}$ data parallels Yen's T_2 results (Fig.2.1) and yields an activation energy of 7370 ± 270 cal/mole. The value and assigned error limit are the results of a reanalysis of the data /Barroilhet, 1977/.

It was proposed that $T_{1\rho}$ at lower temperatures contained a distinguishable component arising from the [129]Xe chemical shift modulated by diffusing vacancies. A slow temperature variation of the data was interpreted as reflecting $E_M - E_F$ = 1060 ± 200 cal/mole where E_M and E_F are the mobility energy and formation energy, respectively, for lattice vacancies. Using E_D = 7370 ± 270 = $E_F - E_M$ Barrhoilet /1977/ concludes that

$$E_M = 4215 \pm 470 \text{ cal/mole} \tag{2.4a}$$

and

$$E_F = 3155 \pm 470 \text{ cal/mole} \quad . \tag{2.4b}$$

If this interpretation is correct, the results may be compared with estimates for xenon by Flynn /1972/, made using corresponding states and krypton data: E_F = 2470 cal/mole and E_M = $2E_F$ = 4940 cal/mole. Also Doyama and Cotterhill /1970/ have calculated E_F = 3800 cal/mole for xenon. The NMR results (2.4) are reasonable, but the preceding analysis is made somewhat uncertain by competing relaxation processes.

[131]Xe. Warren /1966,1967/ reported 3.0 MHz pulsed NMR of [131]Xe in natural xenon. The rigid lattice free induction decay was intermediate between a Gaussian and an exponential and was identified as arising from the central $(-\frac{1}{2} \leftrightarrow \frac{1}{2})$ transition

of the resonance. The central transition is broadened in this case by second order quadrupole interaction with electric field gradients arising from random defects in the cubic xenon lattice. A nearly Gaussian [131]Xe quadrupole echo and reproducibility from one polycrystalline xenon sample to another indicated that the density of defects was large enough that each [131]Xe nucleus was affected by several sources of electric field gradient and that the gradient distribution did not arise from the presence of a hexagonal phase.

The powder-averaged satellite $(\frac{1}{2} \leftrightarrow \frac{3}{2})$ and $(-\frac{1}{2} \leftrightarrow -\frac{3}{2})$ transitions gave rise to a broad resonance line which manifested itself in a sharp quadrupole echo resolvable from the dipolar solid echoes by appropriate rf pulse settings. For a typical xenon polycrystalline sample the ratio of static quadrupolar and dipolar interaction strengths was $\omega_Q/\omega_D \simeq 212$. The $(-\frac{1}{2} \leftrightarrow \frac{1}{2})$ central line became motionally-narrowed above about 110 K. The onset temperature and the temperature dependence of the narrowing are consistent with the coefficient of xenon self-diffusion determined for [129]Xe by Yen /1963/, (2.3).

[131]Xe T_1 data were taken in solid xenon down to 9 K. A log-log plot in Fig.2.2 shows T_1 as a function of T. The solid line indicates the temperature dependence predicted for quadrupole relaxation via a two-phonon Raman process /Van Kranendonk, 1954/. The theoretical curve has been normalized to the data at 77 K and the Debye temperature for solid xenon has been taken to be 55 K.

The Van Kranendonk formalism for Raman scattering of thermal phonons gives rise to first-order spin transition probabilities of the form /Warren, 1966/

$$W_{m,m+N} = C|\langle m + \mu|Q|m\rangle|^2 T*^2 \sum_{n=1}^{6} N_{\mu n}(B\lambda)D_n(T*) \tag{2.5}$$

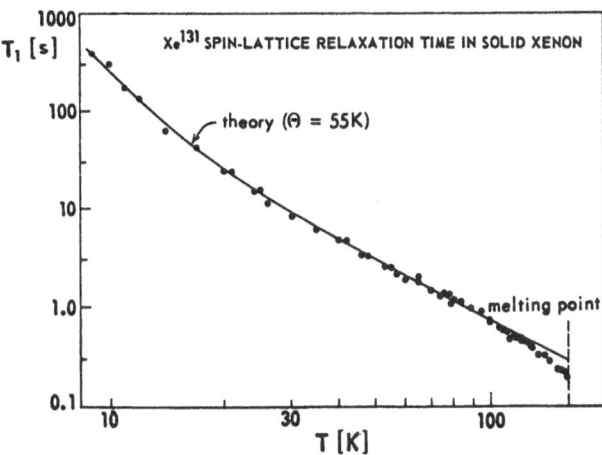

Fig.2.2. Spin-lattice relaxation times of [131]Xe in solid xenon. The solid line represents the temperature dependence of (2.5) normalized to the data at 77 K /Warren, 1966/

where $T^* = T/\Theta$ and $C = 3n_0/\pi v^3 d^2$, with n_0 the number of atoms per unit volume, d the mass density of the crystal, v the velocity of sound, and Θ the Debye temperature. The sum contains products of lattice functions $D_n(T^*)$ determined by the crystal structure and gradient functions $N_{\mu n}(B\lambda)$ which depend on the distribution of charge about the nucleus.

Van Kranendonk /1954/ has shown that in the harmonic phonon case (2.5) gives rise to a temperature dependence of T_1 given by

$$\frac{1}{T_1} \propto T^{*2}E(T^*) \qquad\qquad (2.6)$$

where the function $E(T^*)$ has been evaluated numerically. For $T^* \gg 1$ the theory predicts $1/T_1 \propto T^{*2}$ and for $T^* \ll 1$, $1/T_1 \propto T^{*7}$. The data of Fig.2.2 are seen to lie principally in the high temperature T^{*2} limit. The theory and the temperature dependence of the data near the melting point can be brought into excellent agreement by employing a temperature-dependent Debye temperature, drawn from the heat capacity results of Packard and Swenson /1963/.

In order to estimate the magnitudes of the relaxation rates arising from the harmonic phonon process, Warren /1966/ calculated six unique lattice functions $D_n(T^*)$ for the fcc rare gas solids in a high temperature approximation. The exchange contribution to the gradient functions $N_{\mu n}$ in solid xenon was calculated by assuming that the crystalline electronic wave function is given by the overlap model /Löwdin, 1948/. Exchange — van der Waals cross terms are small /Adrian, 1965/ and the fcc gradient arising from the van der Waals interaction follows from the calculation by Adrian /1965/. Numerical evaluation for solid xenon gave a calculated [131]Xe spin lattice relaxation time at 100 K of $T_1 = 0.3$ s, which compares reasonably well with the 0.7 s observed at that temperature.

Diffusing impurities can provide a significant additional relaxation of [131]Xe in warm solid xenon. A sample with a 10^{-3} air contamination showed an additional exponential decrease of T_1 as the triple point was approached /Warren, 1966/. The relaxation probably arises from quadrupolar interaction via the diffusing impurities. A similar phenomenon has been observed for [83]Kr in impure warm solid krypton /Madaras, 1979/.

[83]Kr. Cowgill /1973,1976/ reported [83]Kr pulsed NMR experiments at 8.475 MHz in solid and liquid krypton samples of normal isotopic abundance. Spin-spin relaxation times in the solid were measured from envelopes of 90°-τ-180°${}_{90°}$ spin echoes. The envelopes showed exponential decay. The two decays were interpreted, respectively, as arising from quadrupole splitting of the spin 9/2 resonance structure and from the $(-\frac{1}{2} \leftrightarrow \frac{1}{2})$ central transition. Quadrupole 2τ echoes were generated with 90°-τ-$\beta_{90°}$ pulse sequences and analyzed to indicate that a typical [83]Kr nucleus is simultaneous-

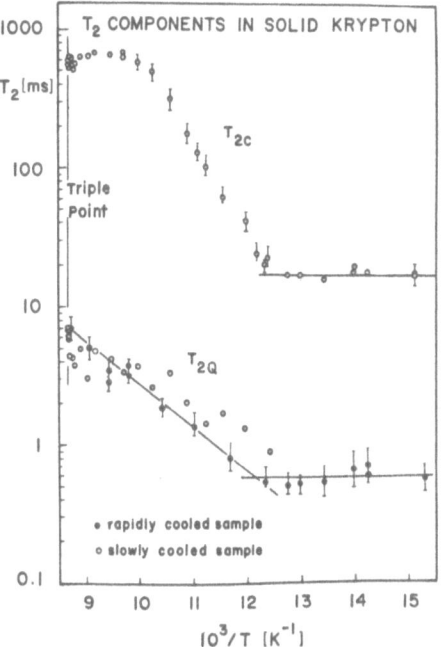

Fig.2.3. Temperature dependence of the central (T₂C) and quadrupolar (T₂Q) components of the spin-spin relaxation in solid krypton /Cowgill, 1976/

ly affected by a large number of lattice defects. The rigid lattice ^{83}Kr resonance line is interpreted to represent a polycrystalline-averaged well-resolved first order quadrupole broadening.

Figure 2.3 shows the observed spin-spin relaxation times T_{2C} and T_{2Q} associated with the central and broad satellite components, respectively. Below 80 K the rigid lattice exponential T_2 components correspond to a static dipolar $\omega_D \equiv 1/T_{2C}^{RL} = 59 \text{ s}^{-1}$, a quadrupolar $\omega_Q \equiv 1/T_{2Q}^{RL} = 1.7 \cdot 10^3 \text{ s}^{-1}$, and thus to the ratio $\omega_Q/\omega_D \simeq 29$. Second order quadrupolar broadening effects were shown to be negligible. Above 80 K both the central line and the satellite component become motionally-narrowed and T_{2C} and T_{2Q} increase with temperature.

Figure 2.4 shows T_{2C} and T_1 in solid krypton over the region of motional narrowing. The solid dots show the ^{83}Kr spin-lattice relaxation results in clean solid krypton above 70 K. The solid line through the T_1 data has been normalized to the data at 80 K and drawn with the temperature dependence predicted by Van Kranendonk and Walker /1967,1968/ for nuclear quadrupole relaxation by anharmonic scattering of thermal phonons. Neglecting the effect of lattice expansion on the exchange contribution to lattice field gradients between 80 and 116 K the anticipated temperature dependence of the relaxation rate can be written

$$\left(\frac{1}{T_1}\right)_{aR} = AT^{*2}E(T^*) \quad . \tag{2.7}$$

65

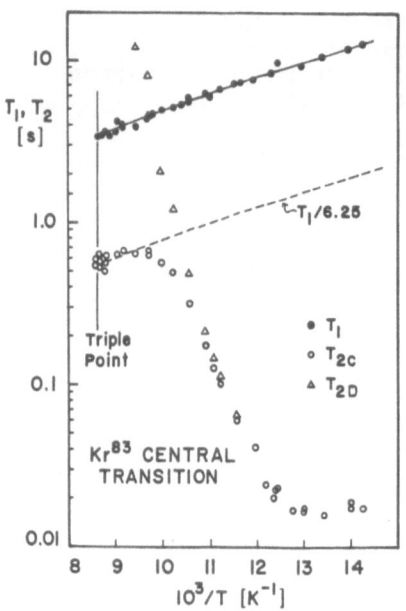

Fig.2.4. T_2 for the central $\frac{1}{2} \leftrightarrow -\frac{1}{2}$ transition and correction for T_1 related cross relaxation in krypton /Cowgill, 1976/

Here the interaction strength A includes γ_G^2 where γ_G is the krypton Grüneisen parameter, which the T_1 data indicate decreases by 12% between 80 and 116 K. Diffusional contributions to dipolar and quadrupolar spin-lattice relaxation are negligible in clean solid krypton samples.

T_{2C} in Fig.2.4 appears to become T_1-limited near the melting point, but at a magnitude substantially smaller than T_1. A calculation by Fedders /1976a/ has shown that, in the presence of a large first-order static quadrupole interaction, a new phenomenon which he has named intraspin cross-relaxation can arise from the thermal phonon-quadrupole mechanism responsible for T_1 in solid krypton. In the absence of other T_2 processes the central $(-\frac{1}{2} \leftrightarrow \frac{1}{2})$ transition for spin I = 9/2 is predicted to be characterized by the ratio $T_1/T_{2C} = 6.25$.

The dashed line in Fig.2.4 corresponds to the solid line through the T_1 data, divided by 6.25. The agreement with the T_{2C} results near the melting point represents a convincing experimental confirmation of the Fedders intraspin cross relaxation effect.

A residual dipolar term T_{2D} was deduced by defining

$$\frac{1}{T_{2D}} \equiv \frac{1}{T_{2C}} - \frac{6.25}{T_1} \quad . \tag{2.8}$$

The corresponding T_{2D} values are indicated as triangles on Fig.2.4. The usual relation between the motionally-narrowed T_{2D} and the coefficient of atomic self-diffu-

sion, (2.1), also is modified in the presence of quadrupolar broadening. Fedders
/1976a/ has shown that the observed dipolar spin-spin relaxation rate for spin 9/2
increased by a factor of 5.96. Thus one has, from (2.1)

$$\frac{1}{T_{2D}} = \frac{5.96}{T_{2Wolf}} = 5.47 \, f \, (2/a_0)^6 \quad . \tag{2.9}$$

Application of (2.9) to the T_{2D} values of Fig.2.4 yielded krypton self-diffusion
results over the $10^3/T$ interval from 9.44 to 11.55. The krypton lattice parameters
of Losee and Simmons /1968a,b/ were employed in the conversion. The results were
(Table 2) $D = 3.1 \, {}^{+6.8}_{-2.1} \, cm^2/s$ and $E_D = 5010 \pm 220$ cal/mole /Cowgill, 1976/. These are
the hopping frequency results of 1.55 MHz measurements of T_{2D} and of $T_{1\rho}$, the spin
lattice relaxation time in the rotating reference frame. The $T_{1\rho}$ results are converted to D values by an analysis similar to that described for ^{21}Ne in the following section.

Table 2. NMR determined self-diffusion parameters in rare gas solids

	D_0 (cm^2/s)	E_D (cal/mole)	Reference
Neon (^{21}Ne)	$0.15 \, {}^{+0.20}_{-0.07}$	944 ± 38	Henry /1972/ corrected by Sirovich /1977/
	$0.15 \, {}^{+0.03}_{-0.02}$	944 ± 6	Sirovich /1977/
Krypton (^{83}Kr)		5250	Cowgill /1971/
	$3.1 \, {}^{+6.8}_{-2.1}$	5010 ± 220	Cowgill /1976/
	$9.0 \, {}^{+1.0}_{-0.6}$	5200 ± 75	Madaras /1981/
Xenon (^{129}Xe)	$9.7 \, {}^{+0.5}_{-0.3}$	7400 ± 50	Yen /1963/, Cowgill /1976/
		7370 ± 270	Barroilhet /1973/

The line drawn in Fig.2.5 is a least square fit of an Arrhenius relation to the
ω_a data. It corresponds to the best present result for atomic self-diffusion in solid krypton

$$D_0 = 9 \, {}^{+1.0}_{-0.6} \, cm^2/s \tag{2.10a}$$

$$E_D = 5200 \pm 75 \text{ cal/mole} \tag{2.10b}$$

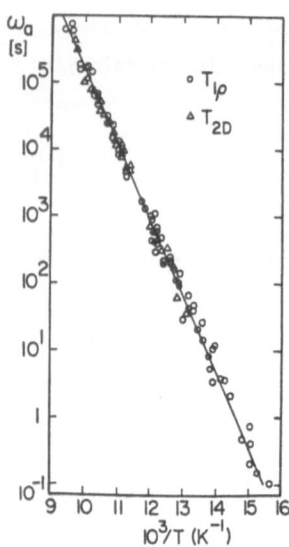

Fig.2.5. Atomic hopping frequency associated with self-diffusion in solid krypton. The line indicated is a least square fit of an Arrhenius relation to the ω_a data from $T_{1\rho}$ (circles) and T_{2D} (triangles) /Madaras, 1983/

which is consistent with the earlier determination by Cowgill /1976/ based on the more limited warm sample T_2 data of Fig.2.4.

Chadwick and Morrison /1968,1970/ reported results of tracer measurements of diffusion in solid krypton

$$D_0 = 5 \, {}^{+5}_{-2} \text{ cm}^2/\text{s} \tag{2.11a}$$

and

$$E_D = 4800 \pm 200 \text{ cal/mole} \quad . \tag{2.11b}$$

Also reported were measurements on a sample containing 0.4% oxygen. These latter results nearly coincide with the NMR diffusion data. However the NMR samples were reported to have had less than 1.7 ppm O_2.

The motional narrowing of the defect-related broad quadrupole-satellite component of the ^{83}Kr resonance is not completely understood. The increase in T_{2Q} with temperature (Fig.2.3) depended on sample history. When the sample was cooled rapidly T_{2Q} followed a thermally-activated behavior with

$$T_{2Q} \propto \exp[-(1400 \pm 200 \text{ cal/mole})/RT] \quad . \tag{2.12}$$

It is interesting to consider a possible thermal vacancy origin for the temperature dependent T_{2Q}. The increase in T_{2Q} begins near 80 K, where Losee and Simmons /1968a,b/ began to find measurable lattice effects of thermal vacancies.

An adiabatic-narrowed quadrupolar interaction with thermal vacancies should correspond to a relaxation rate

$$\frac{1}{T_2} \propto \frac{\sigma_Q^2}{\omega_v} \tag{2.13}$$

where the second moment σ_Q^2 will be proportional to n_v, the number density of vacancies, and ω_v is the characteristic frequency for vacancy diffusion. For small n_v it follows that

$$\omega_v = \frac{\omega_a}{n_v} \tag{2.14}$$

where ω_a, is the characteristic frequency for atomic motion.

With $\omega_a \propto \exp[-(E_F + E_M)/RT]$ and $\omega_v \propto \exp(-E_M/RT)$ one has

$$\frac{1}{T_{2Q}} \propto \frac{n_v}{\omega_v} = \frac{\omega_a}{\omega_v^2} \propto e^{(E_M - E_F)/RT} \quad . \tag{2.15}$$

Losee and Simmons /1968a,b/ have found $E_F = 1780 \pm 200$ cal/mole and (2.10b) gives $E_F + E_M = 5200 \pm 75$ cal/mole. Accepting these values, E_M is about 3420 cal/mole and $E_M - E_F$ about 1640 cal/mole. However Korpiun and Coufal /1971/ have found $E_F = 1985 \pm 200$ cal/mole, which then corresponds to $E_M \simeq 3215$ cal/mole and $E_M - E_F \simeq 1230$ cal/mole.

The solid line through $T_{2Q}(T)$ in Fig.2.3 has an activation energy of 1400 ± 200 cal/mole in possible agreement with (2.15), the mobile thermal vacancy hypothesis and the two $E_M - E_F$ values given above. The more recent measurements by Madaras /1981, 1983/ indicate a T_{2Q} activation energy of $E_M - E_F = 1300 \pm 100$ cal/mole.

When solid krypton samples are cooled slowly, annealing effects on T_{2Q} become apparent (Fig.2.3) with T_{2Q} annealing rates near 1% per hour at 57 K and 0.1% per hour at 41 K /Madaras, 1981/.

^{21}Ne. Henry /1972/ and Sirovich /1977/ reported pulsed 3 MHz NMR experiments on ^{21}Ne in liquid and solid neon. Both experiments employed samples isotopically enriched to about 52% ^{21}Ne. In the rigid polycrystalline solid Henry observed a resonance line which exhibits unresolved first order quadrupole broadening. The rigid lattice gaussian free induction decay corresponds to a T_2^{RL} of 1.17 ms (Fig.2.6) and nearly equal dipolar and quadrupolar contributions to the second moment of the resonance line. Thus $\omega_Q/\omega_D \simeq 1$ and the gaussian decay shape indicates that the static quadrupolar broadening arises from interaction of each spin with many remote lattice defects.

The ^{21}Ne NMR line narrows about 14 K. Above 20 K spin echo envelopes show a resolved fast T_{2Q} component and the slower T_{2C} central component which is plotted in

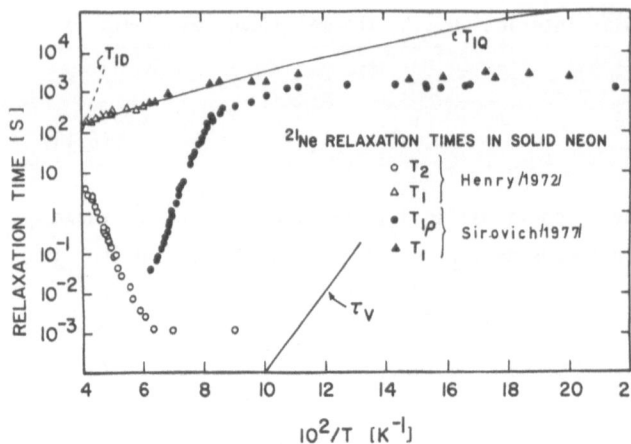

Fig.2.6. ^{21}Ne spin-relaxation times in solid neon. The solid curve indicates the temperature dependence anticipated for quadrupolar T_1. The dashed line indicates a dipolar contribution which has been removed from the observed high-temperature T_1 points plotted. At the bottom a line indicates τ_v /Sirovich, 1977/

Fig.2.6. Near the melting point, ω_c, the characteristic frequency for atomic motion, has become nearly equal to the Larmor frequency ω_0. $T_2(T)$ begins to bend over in an approach to the 10/3 effect and T_1 exhibits a rapidly increasing $1/T_{1D}$ addition to the usual quadrupolar-phonon relaxation rate $1/T_{1Q}$.

Henry /1972/ analyzed the T_{2C} data between 19 and 23 K to yield the Arrhenius self diffusion coefficient (neglecting correlation effects)

$$D_0 = 0.12 \, ^{+0.15}_{-0.05} \, cm^2/s \tag{2.16a}$$

and

$$E_D = 947 \pm 38 \, cal/mole \quad . \tag{2.16b}$$

Sirovich /1977/ reanalyzed these warm solid T_{2C} data to include consideration of thermal expansion of the neon lattice between 19 and 23 K and, via (2.1), to include lattice correlation effects. The correlated results are

$$D_0 = 0.15 \, ^{+0.20}_{-0.07} \, cm^2/s \tag{2.17a}$$

and

$$E_D = 944 \pm 38 \, cal/mole \quad . \tag{2.17b}$$

Both these analyses include a second moment coefficient of 0.85 chosen as a compromise between the "semilike" spin coefficient 0.8 and the "like" spin coefficient 0.9 /Abragam, 1961/.

Sirovich /1977/ extended the NMR investigation of solid neon to lower temperatures. He employed a $\pi/2$ pulse and $\pi/2$ phase shift to achieve a spin lock. This was followed by a programmed adiabatic decrease in the radiofrequency field amplitude to achieve an adiabatic demagnetization in the rotating reference frame (ADRF). In the demagnetized state the spin order of the ^{21}Ne nuclear spins is very sensitive to atomic diffusion /Slichter and Ailion, 1964; Look and Lowe, 1966/. $T_{1\rho}$, the spin-lattice relaxation time in the rotating reference frame, was measured in the strong-collision limit in which the Zeeman, dipolar, and quadrupolar systems establish a common spin temperature between atomic jumps.

Figure 2.6 shows the $T_{1\rho}$ and T_1 results of Sirovich /1977/, together with the earlier high temperature results of Henry /1972/. $T_{1\rho}$ in the thermally activated region was analyzed in the strong-collision approximation via the expression of Wolf /1974b/

$$T_{1\rho} = \tau \; \frac{H_D^2 + H_Q^2 + H_1^2}{0.882 \; H_D^2 + (1-p_Q)H_Q^2} \; . \tag{2.18}$$

Here τ is the characteristic time for atomic diffusion and H_1 is the rotating rf field which, on resonance, is the effective applied field in the rotating frame after demagnetization. H_D and H_Q are dipolar and quadrupolar local fields in the rotating frame. The factor 0.882 is a powder average correlation factor /Wolf, 1974a/ for dipolar local fields in the case of a monovacancy diffusion process on a fcc lattice in the case $\tau_V \ll T_2^{RL} \ll \tau \ll T_1$. Here τ_V is the characteristic time for vacancy motion.

The $T_{1\rho}$ data shown in Fig.2.6 were taken with $H_1 = 0$ and the results between 12 and 15.5 K were analyzed with a least squares fit to an exponential function of reciprocal temperature. The analysis yields an activation energy

$$E_{T_{1\rho}} = 943 \pm 7 \text{ cal/mole} \tag{2.19}$$

which is in agreement with the activation energy, (2.17b), deduced from Henry's earlier T_2 results in warm solid neon. In order to deduce $\tau(T)$ from the $T_{1\rho}(T)$ data one must determine the coefficients in (2.18).

The combined square local field $H_L^2 = H_D^2 + H_Q^2$ can be determined from (2.18) by measuring $T_{1\rho}$ as a function of H_1^2. Such measurements were made at 14.65 K, near the warm end of the $T_{1\rho}$ data. The zero $T_{1\rho}$ intercept yields $H_L^2 = 0.067 \, {}^{+0.014}_{-0.012} \, G^2$.

It can be shown /Hebel, 1963/ that $H_D^2 = \frac{1}{3} <\Delta H_D^2>$ were $<\Delta H_D^2>$ is (in G^2) the Van Vleck /1948/ dipolar second moment of the laboratory frame powder absorption line. A calculation for fcc neon yields $H_D^2 = 0.0280 \, G^2$ and so $H_Q^2 = H_L^2 - H_D^2 = 0.039 \, {}^{+0.014}_{-0.013} \, G^2$ for Sirovich's neon samples.

To complete the solution of (2.18) one either must know τ or $(1-p_Q)$. A $T_{1\rho}/\tau$ fitting was accomplished by extrapolation to 14.65 K of Henry's higher temperature $\tau(T_2)$ results. The same $T_{1\rho}/\tau$ ratio results from fitting all the ln $\tau(T_2)$ and ln $\tau(T_{1\rho})$ points to a single linear function of reciprocal temperature. The ratio also is consistent with the assumption that the static quadrupolar broadening arises from interaction with many randomly distributed remote defects. In this case the correlation factor $(1-p_Q)$ is small and the static electric field gradients are nearly the same at typical adjacent lattice sites. The ADRF $T_{1\rho}$ experimental method excludes signals from [21]Ne nuclei in the immediate vicinity of defects.

The composite set of $\tau(T)$ values are converted to diffusion data using the neon lattice parameters of Schoknecht /1971/. The diffusion results are shown in Fig.2.7. At the upper left are results from Henry's T_2 measurements and at the lower right results from Sirovich's $T_{1\rho}$ data. The straight line is a least squares fit and corresponds to an Arrhenius result for self-diffusion in solid neon between 11.5 and 23 K

$$D_0 = 0.15 \, {}^{+0.03}_{-0.02} \, \text{cm}^2/\text{s} \qquad\qquad (2.20a)$$

and

$$E_D = 944 \pm 6 \text{ cal/mole} \quad . \qquad\qquad (2.20b)$$

The indicated errors include statistical errors and estimated errors in the data points and in the fitting procedures.

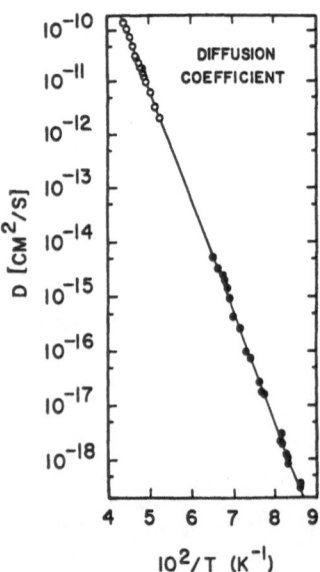

Fig.2.7. Coefficient of atomic self-diffusion in solid neon /Sirovich, 1977/

It should be emphasized that Fig.2.7 and (2.20) are based on several assumptions. The high temperature $D(T_2)$ assume a monovacancy diffusion process and include the choice of a compromise coefficient 0.85 mentioned earlier. The fitting procedure for the low temperature $D(T_{1\rho})$ assumes a monovacancy diffusion process and no off-set between low temperature and high temperature Arrhenius behavior. The assumptions turn out to be consistent with the reasonable result $(1-p_Q) = 0$. Overall it is fair to say that the NMR results of Henry /1972/ and Sirovich /1977/ are consistent with (2.20) and a single Arrhenius monovacancy diffusion process in solid neon between 11.5 and 23 K.

Henry's T_1 data show a dipolar contribution associated with self-diffusion near the melting point. This T_1 component is shown as the dashed line in Fig.2.6 and has been removed from the residual quadrupolar T_1 data plotted as open triangles. The solid line through the T_1 data in Fig.2.6 corresponds to the temperature dependence, (2.7), anticipated for quadrupolar relaxation by Raman scattering of thermal phonons in the Debye approximation.

Below 11 K some additional T_1 mechanism apparently enters and T_1 becomes nearly independent of temperature. The source of the ^{21}Ne spin-lattice relaxation in the low temperature asymptotic region is not clear. Paramagnetic impurities are thought not to be responsible in the case of neon. There is some indication, from the ob-served ratio $T_1/T_{1\rho} \sim 2$, that the relaxation may arise from an independently fluc-tuating quadrupolar field /Fedders, 1976b/. Fedders /1977/ has proposed a novel quad-rupolar spin-lattice relaxation process which may explain the observed ^{21}Ne asymp-totic relaxation. The proposed relaxation arises from an increased phonon spectral density associated with transverse fluctuations of the nuclear spins. It is a weak relaxation mechanism but it is independent of temperature and may explain the ap-proximately 3000 s T_1 values observed in neon. A similar 125 s plateau T_1 is observ-ed for ^{83}Kr in krypton /Madaras, 1981/.

Table 2* summarizes the NMR results for self-diffusion in rare gas solids. The exact agreement between the two sets of neon data is fortuitous. The Madaras results are the best values for krypton. A comparison of many of these NMR results with trac-er measurements, a vacancy model, and corresponding states correlations has been giv-en by Chadwick and Glyde /1977/. Systematic variations among the NMR-determined D_0 and E_D are in good agreement with quantum-corrected corresponding states scaling, Henry /1972/ and Schoknecht /1971/.

Spin lattice relaxation of the quadrupolar nuclides ^{21}Ne, ^{83}Kr, and ^{131}Xe is for the most part well-described by Raman scattering of thermal phonons, (2.5) and (2.6). Figure 2.8 shows a comparison of the predicted and observed temperature variations of T_1 for quadrupolar relaxation of the spin $> \frac{1}{2}$ nuclides in solid neon, krypton, and xenon. The ^{21}Ne plateau may arise from the Fedders /1977/ mechanism described

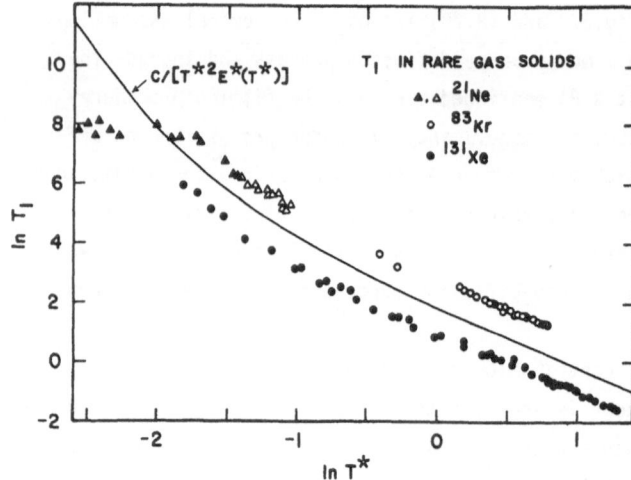

Fig.2.8. T_1 for spin $> \frac{1}{2}$ nuclides in three rare-gas solids as a function of reduced temperature $T^* = T/\Theta_D(T)$. The solid line indicates the temperature dependence of T_1 in (2.6)

earlier. The Debye approximation yields good agreement with the observed temperature variation of T_1, in part because much of the T_1 data lie in or near the universal high temperature T^2 regime /McNeil, 1976/. Also, it is the phonons near the zone boundaries which are most effective in generating fluctuating electric field gradients in the rare gas solids and so the spin relaxation is rather insensitive to details of the true phonon spectrum $G(\omega)$. Nevertheless a correct description of the observed relaxation would entail integration over the full $G(\omega)$. Such an analysis should permit a clearer determination of the degree of anharmonicity present in the relaxation in each case.

2.1.2 Rare Gas Liquids

Diffusion in rare gas liquids. NMR measurements of self-diffusion in liquid neon, krypton, and xenon have been made by observation of spin echoes in experiments which employ pulsed or steady calibrated gradients in the static external magnetic field. The calculated gradients have been calibrated by measuring self-diffusion in a standard fluid (usually water) in the same apparatus. The method is capable of precise results. Any systematic errors usually arise from additional spin-dephasing and thus yield NMR-determined diffusion coefficients larger than the correct values.

Henry's NMR results /1972/ for [21]Ne diffusion in liquid neon are shown as solid dots in Fig.2.9. The measurements were made between 25 and 33 K and at pressures between 0.5 atm and 2.6 atm. These results agree well with [22]Ne tracer measurements at 10 atm by Bewilogua et al. /1971/, shown as open circles in Fig.2.9.

Cowgill's NMR data /1976/ for [83]Kr diffusion in SVP liquid krypton are shown in the upper part of Fig.2.10, together with an indication of the 8.48 atm tracer results reported for [85]Kr by Naghizadeh and Rice /1962/. The line drawn through the

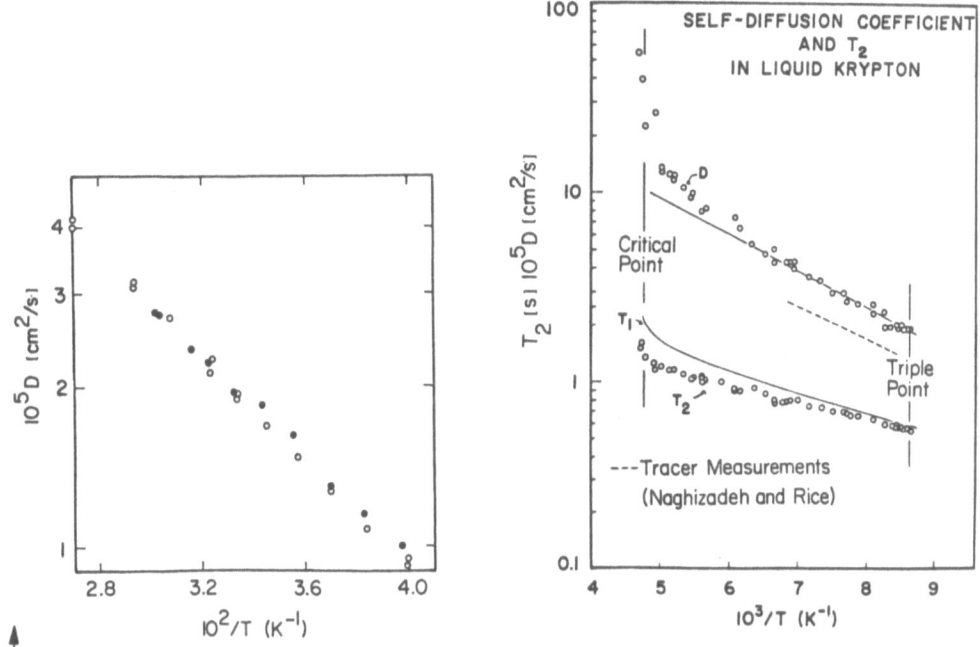

Fig.2.9. Coefficient of self-diffusion in liquid neon. The solid dots are [21]Ne NMR results /Henry, 1972/ and the open circles [22]Ne tracer results /Bewilogua et al., 1971/

Fig.2.10. Temperature dependence of the self-diffusion coefficient of [83]Kr in liquid krypton. T_1 and T_2 results also are indicated /Cowgill, 1976/

NMR data corresponds to a fit of an Arrhenius relation to the results in the cold liquid below 160 K ($10^3/T$ = 6.25). In Fig.2.10 there is a clear systematic deviation from activated Arrhenius behavior for self-diffusion in warm liquid krypton and especially so as the critical temperature in approached.

In warm liquid xenon [129]Xe NMR measurements by Ehrlich and Carr /1970/ showed a similar non-Arrhenius temperature variation of D. In cold liquid xenon earlier [129]Xe NMR measurements by Yen /1963/ and by Hunt and Carr /1963/ and also ^{133}Xe tracer measurements by Naghizadeh and Rice /1962/ indicated nearly Arrhenius behavior with similar activation energies, but with D magnitudes differing by 20 to 80% from the 1970 results.

Ehrlich and Carr /1970/ pointed out that their [129]Xe data along the liquid branch of the coexistence curve can be represented by the empirical relation

$$\rho D = AT^{2.74 \pm 0.08} \quad . \tag{2.21}$$

The fit is illustrated by Fig.2.11. It has been observed that a similar relation can be used to represent diffusion in liquid ethane /Noble and Bloom, 1965/ and methane

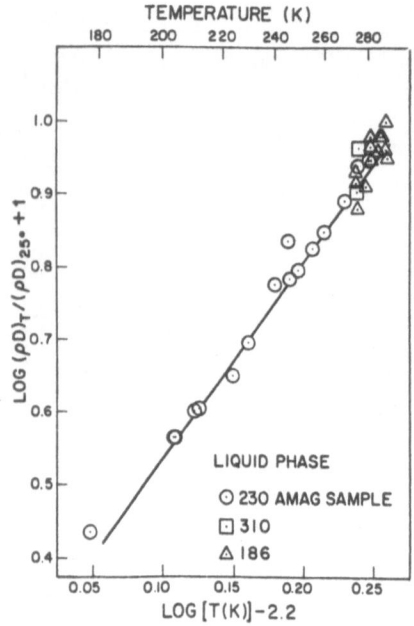

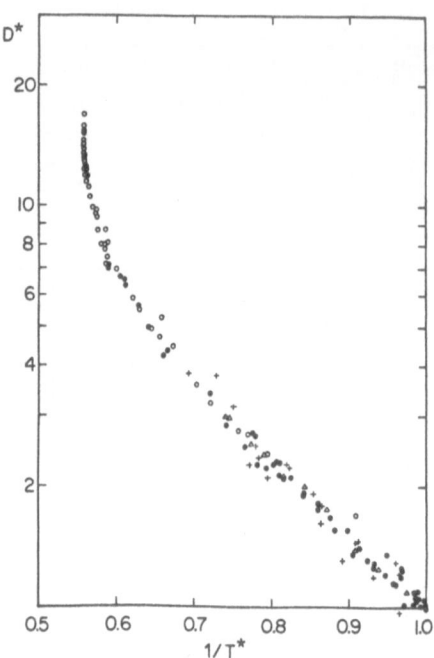

Fig.2.11. The dependence on temperature T of the product of density ρ and the self-diffusion coefficient D on the liquid branch of the coexistence curve for xenon. This product ρD is given relative to its value $(ρD)_{25°C}$ for a sample of near critical density (186 amagat) at 25°C. The straight line is the weighted least-squares fit (2.21) to the data below 281 K /Ehrlich, 1969; Ehrlich and Carr, 1970/

Fig.2.12. Reduced quantities $D/D(T_{Tr})$ vs T_{Tr}/T for rare gas liquids. Open circles are for xenon /Ehrlich and Carr, 1970/; crosses also indicate xenon /Yen, 1962,1963/; solid dots are for krypton /Cowgill, 1976/; triangles are neon /Henry, 1972/

/Ehrlich and Carr, 1970/. (The amagat is a unit of relative density. One amagat corresponds to the gas at 0°C and one atmosphere /Conradi, 1979a/.)

There is a close correspondence among the observed temperature variations of diffusion in the rare gas liquids. Figure 2.12 presents a plot of the reduced quantities $D/D(T_{Tr})$ vs T_{Tr}/T for NMR results in xenon, krypton, and neon. The open circles are for xenon /Ehrlich and Carr, 1970/ and Fig.2.11. The crosses indicate the earlier [129]Xe results of Yen /1962,1963/. The solid dots are the krypton results of Cowgill /1976/ and Fig.2.10. The triangles are the neon results of Henry /1972/ and Fig.2.9. Clearly isotopic diffusion in the rare gas liquids exhibits temperature variations which obey classical corresponding states and which deviate from Arrhenius behavior at temperatures above $T_{Tr}/0.7$ (and perhaps at lower temperatures as well). The power law behavior of the product ρD, (2.21), turns out to work well for all the liquids over the full liquid range (as will be discussed further in Sect.3.1).

Table 3. Self-diffusion parameters for cold rare gas liquids (N = NMR, T = tracer)

	$10^5 D_0$ (cm^2/s)	E_D $(cal/mole)$	$10^5 D(T_{TR})$ (cm^2/s)	p (atm)	Reference
Ne (N)	$66 ^{+21}_{-16}$	211 ± 15	0.87	0.5 to 2.6	Henry /1972/
(T)	84.2 ± 2.8	255 ± 14	0.84 (1.2)	10	Bewilogua et al. /1971/
Ar (T)	116	699	1.7	12.9	Naghizadeh and Rice /1962/
	61	620	1.5 (1.9)	2	Cini-Castagnoli and Ricci /1960/
Kr (N)	103 ± 17	930 ± 40	1.8	SVP	Cowgill /1976/
(T)	48	803	1.45 (1.7)	8.48	Naghizadeh and Rice /1962/
Xe (N)			1.7	SVP	Ehrlich and Carr /1969,1970/
(N)	$230 ^{+80}_{-70}$	1400 ± 150	2.9	10	Yen /1963/
(T)	70	1206	1.6	8.20	Naghizadeh and Rice /1962/

In the vicinity of the triple point the liquid diffusion can be approximated by Arrhenius relations. Table 3 presents a summary of the observed and reduced Arrhenius parameters for self-diffusion in cold rare gas liquids. The tracer measurements of Naghizadeh and Rice /1962/ are restricted to the low temperature liquids ($1/T^* > 0.8$) and yield activation energies somewhat smaller than the NMR results, which have been fitted over a wider temperature range. Some of the data are isobaric and some near SVP, but the pressure-related corrections among the data tabulated are small.

The $D(T_{Tr})$ values indicated in parentheses are the results of the application of the classical theory of corresponding states and based on the xenon results of Ehrlich and Carr /1970/. The NMR determined D for krypton appears to be some 12% too large. The significant deviation for neon probably reflects the effect of quantum mechanical terms needed for correspondence. The deviation from classical correspondence is however much less than that reported for self-diffusion in solid neon /Henry, 1972/.

Spin relaxation in rare gas liquids. Hunt and Carr /1963/ reported T_1 for spin $\frac{1}{2}$ ^{129}Xe in clean liquid xenon to be nearly independent of temperature and about 1000 s between 201 K and 283 K. The relaxation times were independent of external magnetic field between 5 kG and 25 kG and thus the relaxation is not associated with an anisotropic chemical shift mechanism. T_1 probably arises from the mechanism proposed by Torrey /1963/ in which the charge distortion and rotation associated with two collid-ing xenon atoms gives rise to nuclear spin relaxation and a resonant shift (see Sect. 2.2).

In liquid xenon samples containing small amounts of air Yen /1963/ found that the ^{129}Xe T_2 increased abruptly at the melting point and appeared to be T_1-limited near 10 seconds by impurity relaxation in the liquid, Fig.2.1. The liquid relaxation data show the same 1400 cal/mole activated temperature dependence as does the self-diffu-sion in cold liquid xenon. One expects a T_1-limited line with T_2 proportional to D if the correlation times for self-diffusion in the liquid are much less than the Lar-mor periods of the paramagnetic impurities. Hunt and Carr /1963/ reported similar T_1 results in impure liquid xenon.

For all three quadrupolar nuclides ^{21}Ne, ^{83}Kr, and ^{131}Xe T_1 decreases sharply by a factor of about six on melting. Figure 2.13 shows the temperature variation of T_1 for these nuclides. In the rare gas solids, the reduced quantities show the Van Kra-nendonk $E*T*^2$ temperature variation discussed earlier, (2.6) and Figs.2.2,4,6,8. In the cold liquids T_1 increases nearly exponentially as 1/T decreases, but with a tem-perature dependence only about half that of the liquid self diffusion (Fig.2.10). As

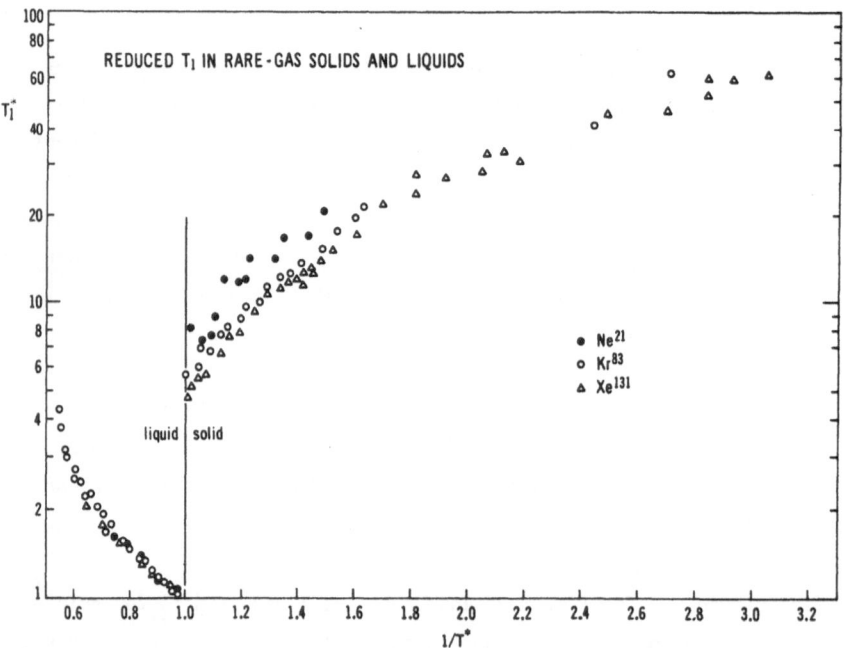

<u>Fig.2.13.</u> Temperature dependence of the liquid and solid T_1 for the rare gas species ^{21}Ne, ^{83}Kr, and ^{131}Xe. Both quantities are reduced by their triple-point values in the liquid /Cowgill, 1973/

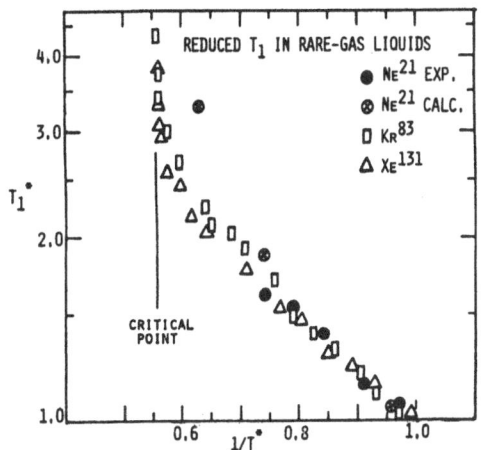

Fig.2.14. Temperature dependence of the liquid T_1 for the rare gas species ^{21}Ne, ^{83}Kr, ^{131}Xe. $T_1^* = T_1/T_1$ (triple point) and $T^* = T/T$ (triple point)/Ringer-macher, 1975/

the critical region is approached, T_1 begins to increase more rapidly with increasing temperature. The spin relaxation in the liquids almost certainly arises from dynamic quadrupole interactions and the correspondence among the three liquids is excellent. Figure 2.14 presents a more complete reduced plot of ln $T_1/T_1(T_{Tr})$ vs T_{Tr}/T in the rare gas liquids.

T_2 is equal to T_1 for ^{131}Xe in liquid xenon. Measurements at 8600 G /Warren, 1966/ and at 51,600 G /Ringermacher, 1975/ show the spin lattice relaxation to be independent of magnetic field. In liquid krypton the temperature variation of the ^{83}Kr T_1 /Cowgill, 1972/ deviates systematically from that of T_2 as the temperature increases, Fig.2.10. The source of the deviation is not presently understood.

Warren /1974/ has described a general theory of nuclear quadrupolar relaxation in extreme-narrowed monatomic liquids. The nuclear spin relaxation rate is expressed in terms of an integral over the dynamic liquid structure factor and a coefficient representing the variation of the local electric field gradient. One can write

$$\frac{1}{T_{1Q}} = \frac{3(2I+3)}{4I^2(2I-1)} \left(\frac{eQ}{\hbar}\right)^2 J(0) \tag{2.22}$$

where the spectral function may be written

$$J(0) = \int_0^\infty dk\, S(k,0)I(k) \tag{2.23}$$

and the integrand factors into an elastic limit dynamic structure factor $S(k,0)$ which describes the liquid dynamics and a weighting term $I(k)$ which is an integral transform of a function determined by the spatial variation of the electric field gradient. Warren used the approximate theory of Adrian /1965/ to describe the electric field gradient in terms of exchange and van der Waals contributions arising from the presence of a neighboring atom.

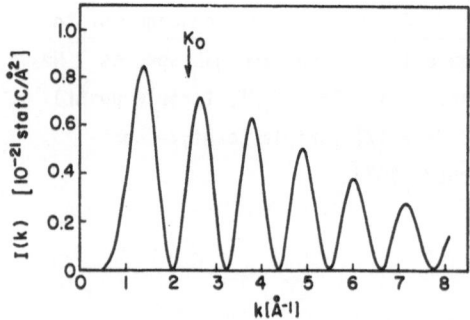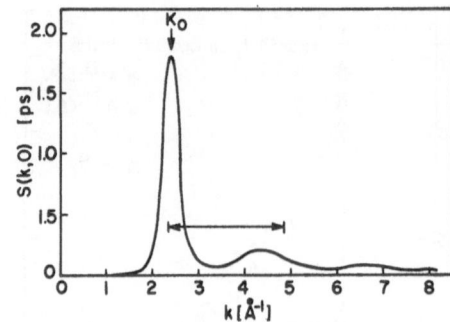

Fig.2.15. a) Exchange-van der Waals EFG weighting function I(k) evaluated for liquid Ne at 25.5 K. Vertical arrow indicates positions (K_0) of main peaks in the liquid structure factor S(k) /Warren, 1974/. b) Elastic ($\omega = 0$) limit of dynamic liquid structure factor S(k,ω) vs k for liquid Ne at 25.5 K. Vertical arrow indicates position of main liquid structure factor peak (K_0). Horizontal arrow indicates range over which relaxation rate integral in (2.23) increases from 10% to 90% of its final value /Warren, 1974/

Warren applied his theory to liquid neon and found that the dominant modes for ^{21}Ne relaxation occur at wave numbers in the range $K_0 \leq k \leq 2K_0$, where K_0 is the position of the main peak in the static liquid structure factor, Figs.2.15a,b. Collective motions of the atoms play a significant role in the ^{21}Ne nuclear relaxation in liquid neon. The calculations neglected quantum corrections, but yielded a relaxation rate at 25.3 K within 25% of that observed by Henry /1972/. The temperature dependence of T_1 predicted by Warren agrees better with the data than does the D/ρ variation which would result from a diffusion model. However there remains a significant deviation between Warren's results and the observations over the limited range from 25.3 to 33.0 K, Fig.2.14. It is possible that further theoretical interpretation of the quadrupolar relaxation data in rare gas liquids can be useful in testing model forms of S(k,0) /Warren, 1974/.

2.2 Chemical Shifts

In the NMR experiments on condensed xenon and krypton it has been found that the magnetic fields measured at the nuclei are shifted appreciably from those in the corresponding dilute gases. The effect arises from interatomic interactions and manifests itself as a density-dependent paramagnetic reduction in the diamagnetic atomic shielding of the atomic electrons. The shift was first reported by Streever and Carr /1961/ and Hunt and Carr /1963/ for ^{129}Xe in xenon liquid and gas and found to be proportional to magnetic field and nearly proportional to sample density. Additional xenon measurements have included solid xenon and both ^{131}Xe and ^{129}Xe: Yen /1963/,

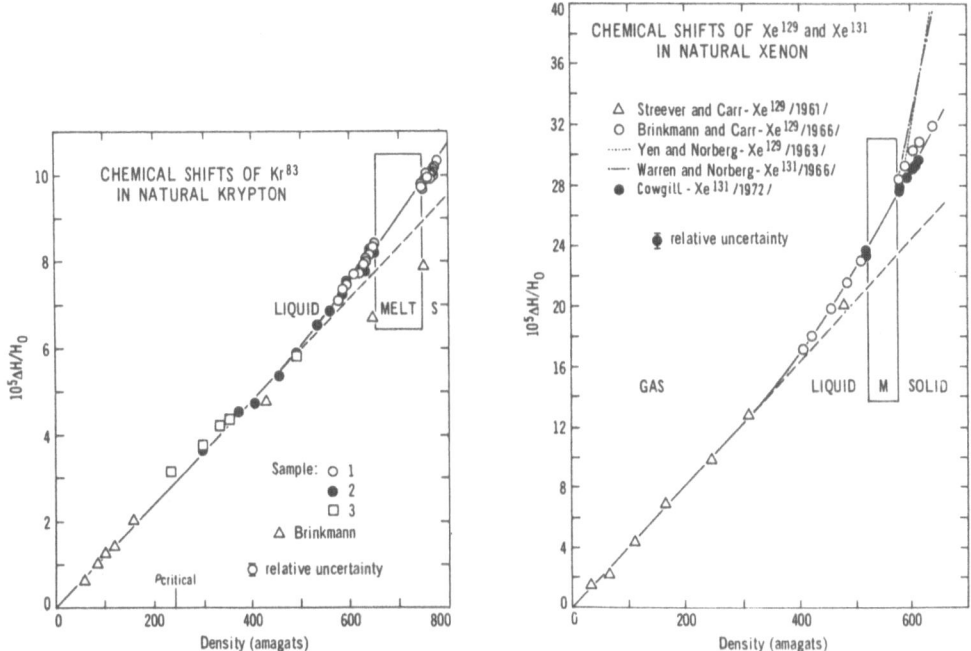

Fig.2.16. Normalized chemical shifts of ^{83}Kr in natural krypton relative to the isolated atom. The solid line represents a smooth fit to the data and indicates their deviation from a linear density dependence (dashed curve) /Cowgill, 1973/

Fig.2.17. Chemical shifts reported in xenon, plotted as a function of density /Cowgill, 1972/

Brinkmann /1964/, Brinkmann and Carr /1966/, Warren /1966/, and Cowgill /1972/. ^{83}Kr shift measurements in condensed krypton have been reported by Brinkmann /1967/ and by Cowgill /1973/.

The experimental results for krypton and xenon are summarized in Figs.2.16 and 2.17. In the gases there are observed resonance shifts ΔH which are approximately linear functions of the density. The solid and liquid data, usually measured as relative shifts as a function of temperature at SVP, show a slight variation from linear density dependence. The experimental results are summarized in Table 4 which lists incremental ratios $\Delta\sigma = (10^7/H_0)(\Delta H/\Delta\rho)$ in units of amagat^{-1} from the data of Figs.2.16 and 2.17.

In liquid and solid krypton the three rather uncertain Brinkmann /1967/ data points, Fig.2.16, lie somewhat below the Cowgill /1973/ results, which have been fitted to the gas phase data. Both sets of data show a continuous density variation of the shift upon melting. Cowgill also reported that a liquid krypton sample contaminated to about 500 ppm with air exhibited shifts 20% larger and with a slightly larger density dependence than observed in a pure sample.

Table 4. Chemical shifts in the rare gases. $(10^7/H_0)(\Delta H/\Delta\rho)$ in units of amagat^{-1} (LJ: Lennard-Jones potential; MS: Munn-Smith potential; MB: Modified-Buckingham potential). From Cowgill /1973/

| | Krypton | | Xenon | |
	Data	Reference	Data	Reference
Gas	1.31 ± 0.15	a	4.22 ± 0.05	b (129)
	0.75	c (theory-LJ)	4.63 ± 0.14	c (129)
	1.77	c (theory-MS)	2.85	d (theory-MB)
Liquid	1.38 to 1.57 with 1.45 average	e	5.66	f (129)
	0.39 ± 0.50	a	5.1 ± 0.5	g (131)
Melting transition	1.54 ± 0.12	e	6.90 ± 0.35	h (131)
	1.0 ± 1.4	a	8.6 ± 0.5	f (129)
			8.1 ± 2.2	g (131)
Solid	1.58 ± 0.24	e	5.72 ± 0.36	h (131)
			3.5 to 8.0 with 6.0 average	f (129)
	1.6 to 0.3 near melting point	i (theory)	4.9	i (theory)

[a] Brinkmann /1967/

[b] Streever and Carr /1961/; Hunt and Carr /1963/

[c] Since gas phase shifts are temperature dependent /Brinkmann, 1963,1968/, the shift magnitudes obtained by this matching procedure are somewhat uncertain.

[d] Adrian /1964,1965/

[e] Cowgill /1973/

[f] Brinkmann /1964/, Brinkmann and Carr /1966/

[g] Warren /1966/

[h] Cowgill /1972/

[i] Lurie and Horton /1966/, Lurie, Feldman, and Horton /1966/

The xenon results shown in Fig.2.17 include measurements made by two research groups and at magnetic fields between 8060 and 51600 G. There is substantial agreement between the data except for two publications which reported shifts in solid xenon with a density dependence three times larger than the data points shown in Fig.2.17. Yen and Norberg /1963/ and Warren and Norberg /1966/ found liquid shifts in agreement with the data points of Fig.2.17, but observed solid shifts which are indicated by the dotted lines in Fig.2.17. Earlier unpublished results obtained by Yen agree very closely with the solid data points of Fig.2.17.

Several attempts were made to resolve the discrepancy among these various results in solid xenon. No appreciable variation of the solid shift was found for xenon samples with different oxygen content, in different sample geometries, or with different thermal histories. Finally the 51600 G measurements of Cowgill and Norberg /1972/ agreed reasonably well with the results of Carr and coworkers. The mystery nevertheless remains undecided. No specific source of error has been identified in the experiment and analyses leading to the anomalous results of Yen, Warren, and Norberg. It is difficult to believe that an allotropic nearly-hcp solid xenon could be responsible for such differing resonance shifts. Measurements as function of pressure or on single crystals might provide an answer.

Calculations of the resonance shifts anticipated in condensed xenon have been reported by Lurie, Feldman, and Horton /1966/, and, for krypton by Lurie and Horton /1966/. The method used is an extension of a calculation by Adrian /1964/ of the shielding in xenon gas. Adrian applied the general theory of chemical shifts to the case of xenon atoms undergoing binary collisions in the gas. He considered the effects of both exchange and van der Waals interactions in deforming the electron clouds during collisions and found that the exchange contribution makes the dominant contribution to the paramagnetic shift in xenon gas. Adrian's result may be written

$$\Delta H = \frac{16\beta^2}{\Delta E} <r^{-3}>_{5p} [S_{\sigma\sigma}+S_{\pi\pi}]^2 H_0 \tag{2.24}$$

where β is the Bohr magneton, ΔE is a suitably chosen average energy of an excited atomic state, r is the distance of the electron from the nucleus, and $S_{\sigma\sigma}$ and $S_{\pi\pi}$ are respectively the overlap integrals between $L_z = 0$ and $L_z = 1$ 5p orbitals centered on the two atoms.

In the rare gas solids it is assumed that two nearest-neighbor atoms may be treated approximately as a diatomic molecule, perturbed by lattice effects. Neglecting correlations between the motions of the nearest neighbor atoms the shift may be written

$$\Delta H(T) = A H_0 \Sigma <sin^2\theta exp[-ZR(T)]> \tag{2.25}$$

where the sum is over nearest neighbors, the brackets represent an average over the canonical ensemble, H_0 is the applied field, R is the separation (static plus displacement) between two atoms, θ is the angle between R and H, and Z and A are parameters. Equation (2.25) is evaluated in the quasi-harmonic approximation /Leibfried and Ludwig /1961/.

The calculated shifts for solid xenon were found to be relatively insensitive to the choice of Lennard-Jones 13-6 or 12-6 potentials, or of a Buckingham 13-6 potential. It was found that a single value of A will nearly account for the magnitudes of the shifts observed in the solid, liquid, and gas phases of xenon (with the exception of the anomalous solid results discussed above). Some comparisons of theory and experiment are given in Table 4.

In krypton the calculated shift magnitudes again agree fairly well with the experimental values obtained by matching gas and liquid phase data. However Brinkmann /1968/ has found a substantial difference between the shifts predicted in gaseous krypton for Lennard-Jones 12-6 and Munn-Smith potentials (Table 4).

3. Dilute H_2 in Neon, Argon, and Krypton

3.1 H_2 in Rare Gas Liquids

Conradi et al. /1979a/ have reported 20 MHz pulsed NMR studies of dilute molecular hydrogen (H_2) in liquid krypton, argon, and neon over nearly the entire liquid temperature intervals, near the saturated vapor pressure (SVP) curves of the host liquids. The NMR detects the ortho-hydrogen (o-H_2) component of the normal hydrogen impurities introduced at concentrations of less than 1%.

3.1.1 Diffusion

Diffusion measurements were accomplished in a calibrated static magnetic field gradient and were stated to have a ±6% uncertainty. Results for diffusion of dilute H_2 in liquid krypton are shown in Fig.3.1. The data are plotted with a reduced abscissa $1/T^* = T_t/T$ where T_t is the triple point temperature of the pure host. Figure 3.1 includes, for comparison, reported self-diffusion results for liquid krypton /Cowgill, 1976; Naghizadeh and Rice, 1962/ previously shown in Fig.2.10. The results for H_2 diffusion in liquid argon and liquid neon were qualitatively similar to that shown for krypton. In the cold liquids both the H_2 diffusion and the self-diffusion approximately obey Arrhenius relations with the same activation energy for the H_2 diffusion and the self-diffusion in each case. However the H_2 diffusion coefficient is systematically larger than the host self-diffusion coefficient.

The impurity diffusion and the self-diffusion may be compared more easily by using the empirical power law [(2.21) and Fig.2.11] described for self-diffusion in xenon

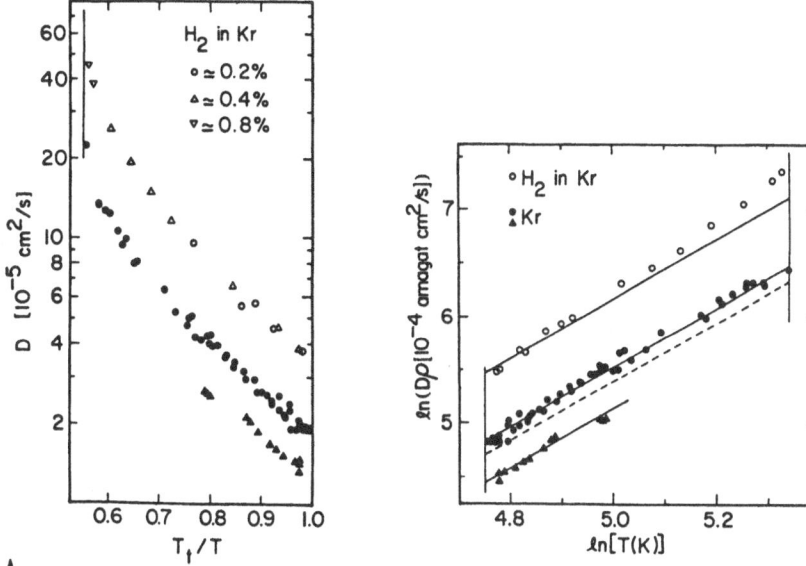

Fig.3.1. Diffusion of dilute H_2 in liquid krypton (open symbols). The solid dots are krypton self-diffusion (NMR, Fig.2.10) and the solid triangles krypton self-diffusion (tracer /Naghizadeh and Rice, 1962; Conradi et al., 1979a/)

Fig.3.2. Temperature variation of $D\rho$ for liquid krypton. Open circles, H_2 in Kr. Solid symbols Kr self-diffusion /Conradi et al., 1979a/

by Ehrlich and Carr /1970/. Figure 3.2 again shows the experimental results for kryp-ton now plotted as ln $10^4 D\rho$ versus ln T, with ρ in amagat for the pure liquid. The triple and critical temperatures are indicated and the sloping straight lines all have been drawn with a slope of 2.74. There is a systematic deviation between the H_2 impurity diffusion and the host self-diffusion in the hot liquid. The dashed line indicates the prediction for krypton self-diffusion of the classical theory of cor-responding states based on the Ehrlich and Carr /1970/ result for self-diffusion in liquid xenon. It appears that the Cowgill /1976/ krypton results may be some 12% too large (Table 3). The NMR H_2-diffusion results disagree by about a factor of two with single points reported from neutron scattering in H_2-Ar and H_2-Ne /Eder et al., 1966/. Tracer results for HT in argon near T_t /Cini-Castagnoli and Ricci, 1960/ lie slightly below the H_2-Ar data, as expected for a heavier mass impurity.

$D\rho$ results for argon and neon are similar to those shown in Fig.3.2. The high tem-perature increases for H_2-Ar and H_2-Ne are systematically larger than those shown for H_2-Kr in Fig.3.2. Figure 3.3 shows, as open symbols and curved lines, the ratios of H_2 diffusion to host self-diffusion in liquid krypton, argon, and neon. Also shown, as solid symbols in Fig.3.3, are calculated ratios interpolated from the results of molecular dynamics computer experiments by Alder et al. /1974/. The computer analyses

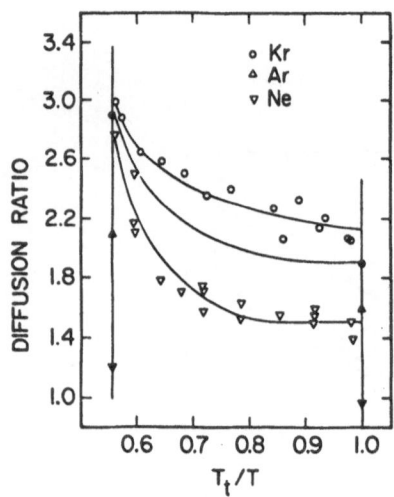

Fig.3.3. Ratio of H_2 diffusion to self-diffusion in liquid neon, argon, and krypton /Conradi et al., 1979a/

were performed for a single impurity in an otherwise pure host fluid with interactions via hard sphere potentials. The poorer agreement between the experimental ratios and the computer results near the critical points is reasonable in view of the larger role of the neglected attractive potential at lower densities. At large densities the diffusion processes are controlled by the hard repulsive forces and it has been shown /Levesque et al., 1970,1973/, that the diffusion coefficient of the Lennard-Jones fluid is within 10% of that for the hard sphere fluid.

3.1.2 Relaxation Times

The proton spin relaxation of dilute o-H_2 in liquid krypton, argon, and neon arises from the intramolecular spin interactions, and the electronic relaxation of the H_2 molecules. These interactions include the proton-proton magnetic dipolar interaction and the spin-rotation interaction involving the net nuclear spin and the rotation of the molecule. The temperatures of the experiments were below 210 K and the o-H_2 restricted (to good approximation) to $J = 1$ rotational states. Transitions among the three m_J sublevels of the $J = 1$ rotational manifold are caused by the relatively weak anisotropic part of the H_2-rare gas intermolecular potential. Relaxation of the molecular spin J is in turn related to the observed relaxation of the protons.

Hardy /1965/ showed that the motion of the molecular spin ($J = 1$) can be described by two correlation times τ_1 and τ_2 which are respectively the correlation times for the first and second order multipole J operators. It can be shown that for many cases, $\tau_1/\tau_2 = 0.6$ for H_2-rare gas atom collisions /Kinsey et al., 1968/. In the extreme-narrowing regime $T_1 = T_2$ and

$$\frac{1}{T_1} = 3.9 \cdot 10^{12} \, \tau_1 \quad , \tag{3.1}$$

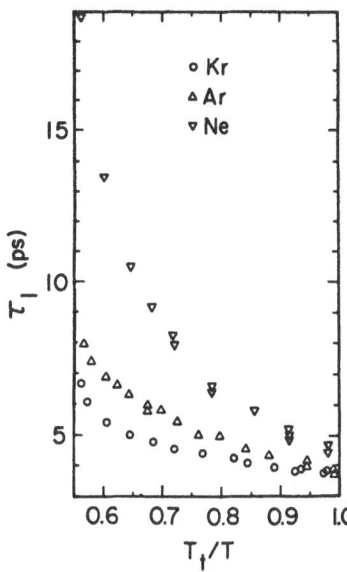

Fig.3.4. Molecular correlation time τ_1 of o-H_2 in liquid neon, argon, and krypton /Conradi et al., 1979a/

where T_1 and τ_1 are measured in seconds. The molecular correlation times τ_1 corresponding to the observed T_1 data for o-H_2 in liquid krypton, argon, and neon are shown in Fig.3.4. As might be anticipated for the nearly spherical H_2, the molecular τ_1 values are considerably longer than the kinetic correlation times of the host rare gas liquids.

3.2 H_2 in Rare Gas Solids

Conradi et al. /1979b/ studied dilute normal-H_2 in solid krypton, argon, and neon. The n-H_2 concentrations ranged from $2.7 \cdot 10^{-5}$ to $8 \cdot 10^{-4}$. Polycrystalline samples were quickly frozen from liquid H_2-rare gas mixtures.

Spin relaxation times T_1 and T_2 were measured for protons in the ortho-H_2 molecules. In the warm solids near their melting points motional narrowing of dipolar broadenings were observed. These yielded information about the inter-diffusion of H_2 and the hosts. Figure 3.5 shows data for warm solid krypton. The solid triangles indicate the residual H_2-Kr intermolecular T_2' obtained after correction of T_2 for limitation by intramolecular T_1 and for H_2-H_2 dipolar broadening. These solid triangle hydrogen T_2' data are to be compared with the open triangles which are ^{83}Kr dipolar central component T_2 values (Fig.2.4) reported by Cowgill /1976/. The horizontal lines correspond to rigid lattice T_2 values of 17 ms for ^{83}Kr and 1.5 ms for o-H_2.

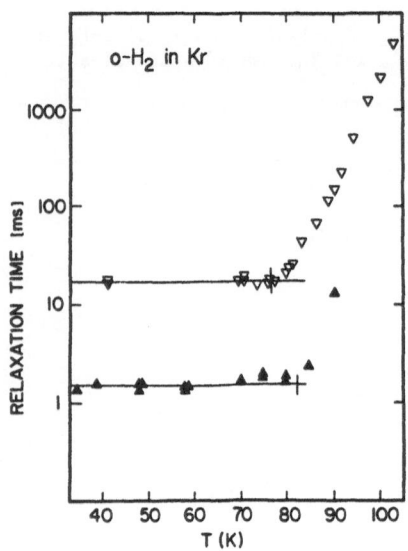

Fig.3.5. Motional narrowing in warm solid krypton. Solid triangles, intermolecular o-H_2 T_2' from H_2-Kr interaction. Open triangles, ^{83}Kr dipolar T_2 from Fig.2.4 /Conradi et al., 1979b/

The motionally-narrowed dipolar T_{2Kr} data for ^{83}Kr were reported to yield the coefficient of self-diffusion in solid krypton given by (2.10). The corresponding correlation time $\tau = 6D/a^2$ (where a is the lattice nearest neighbor separation) is 17 ms at 76.8 K and 1.5 ms at 82.6 K (indicated by vertical lines in Fig.3.5). The correspondence between the two intersections and the observed onset of motional narrowings of the ^{83}Kr and the o-H_2 rigid lattice lines indicates that the coefficient of H_2-Kr diffusion near 80 K is about the same as that reported for Kr self-diffusion.

Similar analyses of the narrowing of H_2-H_2 intermolecular dipolar broadenings for H_2-Ne and H_2-Ar indicate that in the warm solids the H_2 diffusion coefficients again are about the same as those for self-diffusion in solid neon and solid argon. However these results must be regarded as uncertain because of the limited data range and because the coefficient of self-diffusion in argon is not known to good accuracy /Chadwick and Glyde, 1977/.

At the low H_2 concentration studies, the proton spin-lattice relaxation in the rare gas solids (as in the liquids) was dominated by intramolecular spin interactions and the electronic relaxation of the ortho-H_2 molecules which were restricted to the J = 1 manifold. Thus, as for H_2 in the rare gas liquids, spin relaxation is related to correlation times τ_1 and τ_2 for the first and second order multipole J operators. In discussing relaxation in the solids we shall consider the corresponding correlation frequencies or decay rates Γ_1 and Γ_2.

Because the same intramolecular spin interactions are present for o-H_2 in H_2 gas and in rare gas solids the T_1 (Γ_1, Γ_2) might be expected to be the same in the two systems. However Fedders /1979/ has shown how static crystal fields influence the relaxation of H_2 in solids and that measurements of T_1 and T_2 can yield definite in-formation about the environment of H_2 molecules in a host lattice. Large crystal fields shift certain terms in the multipole expansion of the intramolecular spin interactions to very high frequencies. At these frequencies, much larger than the Larmor frequency, the terms do not contribute to relaxation. Fedders has calculated T_1 under a variety of assumptions about the static crystalline electric field gra-dients. For cubic symmetry, i.e., no gradients, Fedders obtains the same result as for H_2 gas. For a "no symmetry" case where there are field gradients with neither cubic nor axial symmetry Fedders predicts that the minimum average T_1 for H_2 in a polycrystalline sample will be between 0.534 ms ($r = -1$) and 0.587 ms ($r = +1$) at 20 MHz. The observed T_1 minima for H_2 in solid neon and argon lie in this range. The parameter r describes the relative magnitudes of transverse components V_{xy}^2 and $(V_{xx} - V_{yy})^2$ of the electric field gradient tensor. For axial symmetry the predicted minimum average T_1 is 0.285 ms.

For example Fedders' result for the longitudinal decay rate, averaged over all angles for a polycrystalline sample in the "no symmetry" case is

$$\frac{1}{T_1} = \frac{2}{5} \frac{\omega_d^2}{\omega_0} \left[\frac{(3-r)x}{1+x^2} + \frac{(12+r)x}{4+x^2} \right] \quad . \tag{3.2}$$

Here $x = \Gamma_2/\omega_0$, ω_0 is the nuclear Larmor frequency, ω_d is $3.62 \cdot 10^5$ s^{-1} for H_2 /Abra-gam, 1961/, and Γ_2 is the decay rate (correlation frequency) characteristic for the quadrupole correlation functions.

The temperature variation of T_1 thus reflects the temperature variation of the molecular correlation frequency Γ_2. One way to construct a theory of $T_1(T)$ for H_2 in rare-gas solids is to use the Van Kranendonk model /1954/ and assume that the temperature dependence of Γ_2 is given by

$$\Gamma_2 = CT*^2 E*(T*) \tag{3.3}$$

which was derived using the Debye spectrum of phonons and other approximations. Here we take $T* = T/\Theta_c$ where Θ_c is a characteristic temperature for the system. $E*(T*)$ is the tabulated Van Kranendonk function /1954/, which is unity for temperatures much above Θ_c and is proportional to $T*^5$ for temperatures much less than Θ_c.

Figure 3.6 shows Conradi's average T_1 and T_2 results for dilute H_2 in solid neon. The T_2 points, especially above the T_1 minimum, have been corrected for static broad-enings. The raw T_2, indicated by crosses, show the motional narrowing of intermolecu-lar contributions referred to in the discussion of H_2-Ne interdiffusion, above. The lines drawn indicate the powder average values T_1 and T_2 predicted by Fedders' the-

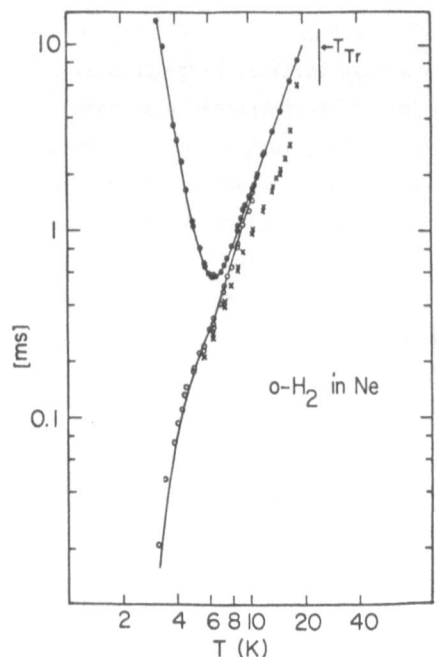

Fig.3.6. T_1 and T_2 for H_2 in solid neon. Crosses indicate T_2, open circles intramolecular T_2, solid dots T_1. The lines indicate the /Fedders, 1979/ theoretical intramolecular T_1 and T_2 for $r = 0$, $T_{(min)} = 6.5$ K, and $\Theta_c = 40$ K /Conradi et al., 1979b/

ory for the "no symmetry" case with $r = 0$. The T_1 minimum has been taken to occur at 6.5 K and Θ_c has been taken to be 40 K. A variation of ± 2 in Θ_c degrades the fit appreciably. There is a similar excellent agreement between theory and experiment for H_2 in solid argon. The conclusion is that H_2 in neon and argon either reside preferentially on sites of low symmetry or else themselves produce local lattice distortions of low symmetry. There were small ($\sim 8\%$) annealing effects towards higher symmetry visible in T_1 for H_2 in solid neon which had been warmed to 18 K for one hour.

Table 5 lists the characteristic temperatures needed in order to fit the combined Van Kranendonk and Fedders models to the T_1 and T_2 data, including an axial symmetry fit for dilute o-H_2 in solid p-H_2. Also shown are average Debye temperatures for the host solids in the temperature region of the NMR data. The lack of variation of the Θ_c parameter is suggestive of a local mode phenomenon for the spin relaxation of H_2 in the rare gas solids.

Relaxation data for H_2-Kr were more complex than for H_2-Ne or H_2-Ar and may have shown an admixture of o-H_2 sites with axial symmetry /Meyer et al., 1954/ or other higher symmetry and also a significant contribution from trapped pockets of H_2 liquid and gas.

The Fedders theory predicts an unusual correlation in that, at low temperatures, a region of a solid sample which contributes a short T_2 component also will have a

Table 5. Parameters for molecular decay rates Γ_2 of dilute o-H_2 in various solids. $\Gamma_2 = CT*E*(T*)$. $T* = T/\Theta_c$

	Neon	Argon	Para-H_2
Θ_c (K)	40	40	43
Θ_D (K)	66[a]	81[b]	100[c]
C (10^9 s^{-1})	38	3.1	2.2

[a]Korpiun and Lüscher /1977/

[b]Klein and Koehler /1976/

[c]Wert and Thomson /1970/

long T_1 component and vice-versa. In general a relation T_1T_2 = constant holds at temperatures well below the T_1 minimum. Conradi observed that for H_2-Ar the low temperature T_1 recoveries and T_2 decays were significantly nonexponential and varied from one sample to another. Nevertheless if the relaxation processes were analyzed into fast and slow components then the products ($T_{2short} T_{1long}$) and ($T_{2long} T_{1short}$) were constant, independent of temperature for a given sample.

The smooth lines drawn in Fig.3.6 are the result of introducing into the Fedders theory a decay rate $\Gamma_2(T) = CT*^2E*(T*)$ from the Van Kranendonk model. However the Fedders theory does not require employment of such a model. For the "no symmetry" case for the H_2 sites in Ne and Ar (3.2) can be used to convert the T_1 data (and a similar relation the T_2 data) directly to deduced values of the molecular correlation frequency Γ_2. Figure 3.7 shows the experimental relaxation results for dilute H_2 in solid neon and argon, converted to $\Gamma_2(T)$. Also shown are $\Gamma_2(T)$ for dilute o-H_2 in solid p-H_2 analyzed via axial symmetry for the local field gradients. The curved lines indicate the $\Gamma_2(T)$ from the Van Krankendonk model and (3.3) used earlier in fitting the data (cf. Fig.3.6). At the upper right are indicated the correlation frequencies $\Gamma_2 = (\tau_2)^{-1}$ deduced via (3.1) from the relaxation data for liquid mixtures (and shown as correlation times τ_1) in Fig.3.4.

One sees that (3.3) and a characteristic temperature near 40 K gives an excellent fit to the $\Gamma_2(T)$ results. Further, the value of Γ_2 is about the same for H_2 at the melting points of solid neon, argon, and krypton. The $\Gamma_2(T)$ values for o-H_2 in the solids range from $4 \cdot 10^6$ to $1.2 \cdot 10^{10}$ s^{-1}. The data pass smoothly through the $1.9 \cdot 10^8$ s^{-1} value for the $r = 0$ no symmetry T_1 minimum, which behavior gives further evidence of the quality of the description provided by the Fedders theory.

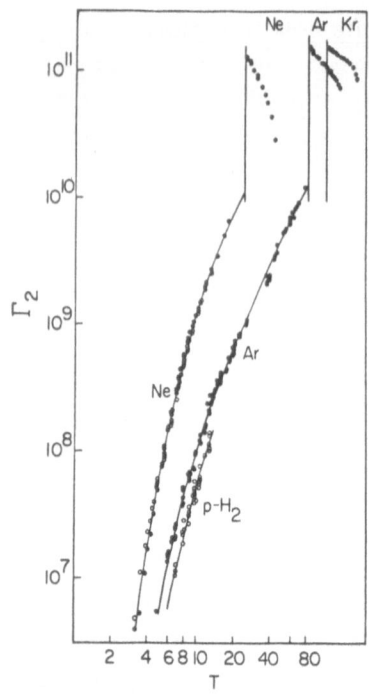

Fig.3.7. Molecular correlation frequencies Γ_2 for dilute o-H_2 in solid neon, argon, and p-H_2. For neon the open circles are derived from T_2 and the solid dots from T_1. At the upper right are Γ_2 for o-H_2 in liquid neon, argon, and krypton /Conradi et al., 1979b/

It is remarkable that the molecular correlation frequency Γ_2 for o-H_2 is about the same at the triple points of the rare gas solid hosts investigated, and that Γ_2 increases by a factor of about 12 upon melting.

4. Concluding Remarks

NMR results have been summarized for magnetic nuclides in condensed neon, krypton, and xenon. Quadrupolar spin-lattice relaxation reflects Raman scattering of phonons in the solids and a degree of collective excitation in the liquids. Self-diffusion results in rare gas solids now have a particularly high accuracy because of the inclusion of studies of ultra-slow atomic motions. The NMR results in krypton provide a vacancy formation energy significantly smaller than those predicted by theories, which apparently do not adequately describe the effects of manybody forces in krypton and xenon. This review has not discussed several aspects of the results of Madaras /1981,1983/ on solid krypton. These include measurements of the field dependence of spectral diffusion in the quadrupole-broadened spin 9/2 [83]Kr resonance line.

A number of advances have arisen from the work of Fedders /1976,1977,1979/ on the effects of electric field gradients on spin decay rates. Some of these include: the effects of intraspin cross-relaxation (the mixing of spin decay rates) on line shapes and on central component $\frac{1}{2} \rightarrow -\frac{1}{2}$ decay rates, the field dependence of relaxation times T_1 and $T_{1\rho}$, relaxation effects of atomic vacancy hopping, effects of self-consistency on spin-lattice relaxation, and motion-induced relaxation in the low temperature-high frequency limit. Each of these advances in theoretical understanding, and others now in progress, provides opportunities for new or improved quantitative measures of structure and dynamics in solids, and particularly in rare gas solids, where there is a large useful range of quadrupole interaction strengths.

NMR studies of o-H_2 as a dilute impurity in condensed neon, argon, and krypton yield diffusion coefficients and also molecular correlation rates for the multipole J operators of ortho-H_2. The o-H_2 relaxation is a sensitive measure of the symmetry of the crystal fields at the H_2 sites. The remarkable temperature Θ_c = 40 K characterizes the local phonon dynamics as they enter the Raman mechanism responsible for the o-H_2 relaxation in rare gas solids. The precise origin of Θ_c is not understood. Experiments currently are in progress to determine whether the relaxation of p-D_2 in rare gas solids also displays Θ_c = 40 K.

Acknowledgments. Much of the work reviewed here has been supported under various National Science Foundation grants, including DMR 80-10818 and 83-04473 and by an Equipment Laon Contract from the Office of Naval Research. The review has benefitted from stimulating discussions with P.A. Fedders.

References

Abragam, A. /1961/: *The Principles of Nuclear Magnetism* (Oxford University Press, London) (Sects.2.1; 3.2)
Adrian, F.J. /1964/: Phys. Rev. *136*, A980 (Sects.2.1,2)
Adrian, F.J. /1965/: Phys. Rev. *138*, A403 (Sects.2.1,2)
Alder, B.J., W.E. Alley, J.H. Dymond /1974/: J. Chem. Phys. *61*, 1415 (Sect.3.1)
Anderson, P.W., P.R. Weiss /1953/: Rev. Mod. Phys. *25*, 269 (Sect.2.1)

Barroilhet, L.E. /1973/: Ph.D. Thesis, University of Wisconsin (Sect.2.1)
Barroilhet, L.E. /1977/: Private communication (Sect.2.1)
Bewilogua, L., C. Gladun, B. Kubsch /1971/: J. Low Temp. Phys. *4*, 299 (Sect.2.1)
Brinkmann, D. /1963/: Helv. Phys. Acta *36*, 413 (Sect.2.2)
Brinkmann, D. /1964/: Phys. Rev. Lett. *13*, 187 (Sect.2.2)
Brinkmann, D. /1967/: Phys. Lett. *25A*, 520 (Sect.2.2)
Brinkmann, D. /1968/: Helv. Phys. Acta *41*, 367 (Sect.2.2)
Brinkmann, D., H.Y. Carr /1966/: Phys. Rev. *150*, 174 (Sect.2.2)

Carr, H.Y., E.M. Purcell /1954/: Phys. Rev. *94*, 630 (Sect.2.1)
Chadwick, A.V., H.R. Glyde /1977/: In *Rare Gas Solids*, ed. by M.L. Klein and

J.A. Venables (Academic, London) Vol. II, p. 1151-1229 (Sects.2.1; 3.2)
Chadwick, A.V., J.A. Morrison /1968/: Phys. Rev. Lett. *21*, 1803 (Sect.2.1)
Chadwick, A.V., J.A. Morrison /1970/: Phys. Rev. B *1*, 2748 (Sect.2.1)
Cini-Castagnoli, G., F.P. Ricci /1960/: Nuovo Cimento *15*, 795 (Sect.2.1)
Conradi, M.S., K. Luszczynski, R.E. Norberg /1979a/: Phys. Rev. B *19*, 20
   (Sects.2.1,2; 3.1)
Conradi, M.S., K. Luszczynski, R.E. Norberg /1979b/: Phys. Rev. B *20*, 2594 (Sect.3.2)
Cowgill, D.F. /1971/: Ph.D. Thesis, Washington University (Sect.2.1)
Cowgill, D.F., R.E. Norberg /1972/: Phys. Rev. B *6*, 1636 (Sect.2.2)
Cowgill, D.F., R.E. Norberg /1973/: Phys. Rev. B *11*, 4966 (Sects.2.1,2)
Cowgill, D.F., R.E. Norberg /1976/: Phys. Rev. B *13*, 2773 (Sects.2.1; 3.1,2)

Doyama, M., R.M.J.Cotterhill /1970/: Phys. Rev. B *1*, 832 (Sect.2.1)

Eder, O.J., S.H. Chen, P.A. Egelstaff /1966/: Proc. Phys. Soc. *89*, 833 (Sect.3.1)
Ehrlich, R.S. /1969/: Ph.D. Thesis, Rutgers University (Sect.2.1)
Ehrlich, R.S., H.Y. Carr /1970/: Phys. Rev. Lett. *25*, 341 (Sects.2.1; 3.1)

Fedders, P.A. /1976a/: Phys. Rev. B *13*, 2768 (Sects.2.1; 4)
Fedders, P.A. /1976b/: Phys. Rev. B *14*, 1842 (Sects.2.1; 4)
Fedders, P.A. /1977/: Phys. Rev. B *15*, 3297 (Sects.2.1; 4)
Fedders, P.A. /1979/: Phys. Rev. B *20*, 2588 (Sects.3.2; 4)
Flynn, C.P. /1972/: *Point Defects and Diffusion* (Clarendon, Oxford) (Sect.2.1)

Hardy, W.N. /1965/: Can J. Phys. *44*, 265 (Sect.3.1)
Hebel, L.C. /1963/: In *Solid State Physics*, ed. by F. Seitz and D. Turnbull
   (Academic Press, New York and London) Vol. 15 (Sect.2.1)
Henry, R., R.E. Norberg /1972/: Phys. Rev. B *6*, 1645 (Sect.2.1)
Hunt, E.R., H.Y. Carr /1963/: Phys. Rev. *130*, 2302 (Sects.2.1,2)

Kinsey, J.K., J.W. Riehl, J.S. Waugh /1968/: J. Chem. Phys. *49*, 5269 (Sect.3.1)
Klein, M.L., T.R. Koehler /1976/: In *Rare Gas Solids*, ed. by M.L. Klein and
   J.A. Venables (Academic Press, London) Vol. I, p. 301 (Sect.3.2)
Korpiun, P., H.J. Coufal: Phys. Stat. Sol. (a) *6*, 187 (Sect.2.1)
Korpiun, P., E. Lüscher /1977/: In *Rare Gas Solids*, ed. by M.L. Klein and
   J.A. Venables (Academic Press, London) Vol. II, p. 754 (Sect.3.2)

Leibfried, G., W. Ludwig /1961/: Sol. State Phys. *12*, 275 (Sect.2.2)
Levesque, D., L. Verlet /1970/: Phys. Rev. A *2*, 2514 (Sect.3.1)
Levesque, D., L.L. Lee /1973/: Mol. Phys. *26*, 1351 (Sect.3.1)
Look, D.C., I.J. Lowe /1966/: J. Chem. Phys. *44*, 2995 (Sect.2.1)
Losee, D.L., R.O. Simmons /1968a/: Phys. Rev. *172*, 934 (Sect.2.1)
Losee, D.L., R.O. Simmons /1968b/: Phys. Rev. *172*, 944 (Sect.2.1)
Löwdin, P.O. /1948/: Ark. Mat. Astron. Fisik *35A*, No. 9 (Sect.2.1)
Lurie, J., G.K. Horton /1966/: Phys. Lett. *22*, 560 (Sect.2.2)
Lurie, J., J.L. Feldman, G.K. Horton /1966/: Phys. Rev. *150*, 180 (Sect.2.2)

Madaras, E.M., R.E. Norberg /1979/: Bull. Am. Phys. Soc. *24*, 490 (Sect.2.1)
Madaras, E.M. /1981/: Ph.D. Thesis, Washington University (Sects.2.1; 4)
Madaras, E.M., R.E. Norberg /1983/: Submitted to Phys. Rev. B (Sects.2.1; 4)
McNeil, J.A. /1976/: Phys. Rev. B *13*, 4714 (Sect.2.1)
Meyer, L., C.S. Barrett, P. Haasen /1964/: J. Chem. Phys. *40*, 2744 (Sect.3.2)

Naghizadeh, J., S.A. Rice /1962/: J. Chem. Phys. *36*, 2710 (Sects.2.1,2; 3.1)
Noble, J.D., M. Bloom /1965/: Phys. Rev. Lett. *14*, 250 (Sect.2.1)

Packard, J.R., C.A. Swenson /1963/: J. Phys. Chem. Solids *24*, 1405 (Sect.2.1)

Ringermacher, H.R., R.E. Norberg /1975/: Proc. 14th Int. Conf. on Low Temp. Phys.,
   Vol. 4 (Elsevier, New York) p. 400 (Sect.2.1)

Schoknecht, W.E. /1971/: Ph.D. Thesis, University of Illinois (Sect.2.1)
Sirovich, B.E., R.E. Norberg /1977/: Phys. Rev. B *15*, 5107 (Sect.2.1)

Slichter, C.P., D. Ailion /1964/: Phys. Rev. *135*, A1099 (Sect.2.1)
Streever, R.L., H.Y. Carr /1961/: Phys. Rev. *121*, 20 (Sect.2.2)

Torrey, H.C. /1963/: Phys. Rev. *130*, 2306 (Sect.2.1)

Van Kranendonk, J. /1954/: Physica *20*, 781 (Sects.2.1; 3.2)
Van Kranendonk, J., M.B. Walker /1967/: Phys. Rev. Lett. *18*, 701 (Sect.2.1)
Van Kranendonk, J., M.B. Walker /1968/: Can. J. Phys. *46*, 2441 (Sect.2.1)
Van Vleck, J.H. /1948/: Phys. Rev. *74*, 1168 (Sect.2.1)

Warren, W.W., Jr., R.E. Norberg /1966/: Phys. Rev. *148*, 402 (Sects.2.1,2)
Warren, W.W., Jr., R.E. Norberg /1967/: Phys. Rev. *154*, 277 (Sect.2.1)
Warren, W.W., Jr. /1974/: Phys. Rev. A *10*, 657 (Sects.2.1,2)
Wert, C.A., R.M. Thomson /1970/: *Physics of Solids* (McGraw-Hill, New York) (Sect.3.2)
Wolf, D. /1974a/: Phys. Rev. B *10*, 2710 (Sect.2.1).
Wolf, D. /1974b/: Phys. Rev. B *10*, 2724 (Sect.2.1)

Yen, W.M.S. /1962/: Ph.D. Thesis, Washington University (Sect.2.1)
Yen, W.M.S., R.E. Norberg /1963/: Phys. Rev. *131*, 269 (Sects.2.1,2)

Combined Subject Index

Activation energy 61,62,67-73,77

ADRF 71,72

After effects in nuclear decay 8,23

Aggregation 6,12

Aluminium 33,45,46,54

Amagat 78,81

Ammonia 43

Anharmonicity in the phonon spectrum 30

Annealing 13,14,39,40,47,50,69,90

Argon crystal 9,14

Arsen 44,53

  hydride 44

Beta decay 8,9

Cadmium 44,45

Central transition 63-66,87

Cesium 40

Change nuclear radius 26,27

Characteristic temperature 89-91

Charge transfer 8,24

Chemical shifts 62,78,80-84

Chromium 44,45

Clustering 6,10,13,14

Cobalt-57, decay 8,23

Conversion, internal 8

Copper 40,41

Correlation factor 71

Correlation times 61,87

Corresponding states 62,73

Crucible 10,12

Cryostat, liquid helium 10,11,15

Crystal field parameter (splitting) 3, 19,23,28,31,32

Debye, frequency 31

Debye-Waller temperature 29,30

Defects 63,68

Deposition rate 6,12,14

Deuterium 37

Dipolar interaction 61-67,70-73,86,87

Dirac-Fock-SCF calculation 26,27

Echo 60,63,69,74

Electric field gradient tensor 4

  with axial symmetry 8,21,22,48

Electron

  acceptor 44

  affinity 8

  capture nuclear decay 8,23

  charge density at the nucleus 2,4, 26,27

  donor 44

Europium 19,20,25,27,30,32

Excited state

  electronic 8,24,48

  nuclear 5,9,16,17,19,21

Gallium 33,46,47,54

Gas handling system 54

g factor 4,20,27

  effective 4,27,28,33,34,45

  shift 5,34,36-41,43,44,52,54

Gold 40,41

g-tensor components 46-48

Hydrogen 7,21,35-37,52-54,84-92
Hyperfine
  coupling constant 4,27,28,33,34,45
  shift 5,34,36-41,43,44,52
  tensor components 46-48

Impurities 61,62,64,78
  residual gas 10-12
Impurity diffusion 84-88
Induction decay 60,62,69
Intraspin cross-relaxation 66
Iodine 8,22,33,48
Ion
  accelerator 9
  implantation 9,10,24
Irradiation
  gamma 1,7,15,35,36,42-44
  infrared 14,39
  neutron 9,14,35
  ultraviolet 7,8,21,22,36,42-44,48
Iron
  atom 17-19, 25-32
  ion 21-32
Isomer shift 4,9,25-27

Jahn-Teller coupling 54

K-band 16
Knudsen cell 6,7
^{83}Kr 60,64-67,74-82
Kramer's doublet 19,23,28,31

Lamb-Mössbauer factor 2,11,29-31
Lattice site *see also* Trapping site
  4,5,11,19,32
Lifetime
  of electronic states 9
  of nuclear states 8,9,16,17,19,21
Linewidth
  in ESR experiments 16,37,42
  in NGR experiments 12,17,18,23
Liquids 74,84-87

Lithium 37,54

Manganese 44,45
Magnetic field
  effective 4,18,27,28
  external 3-5,18,23,27,28,32,33,49
Melting point 13,40
Microwave
  cavity 13,15,16
  power 16
Molecular dynamics 85,86
Mössbauer
  absorber 6,10-12
  source 6,9-11
  temperature 29,30
Motional narrowing 60,61,65-69,88
Muonic isomer shift 27

^{21}Ne 60,67-80
Neutron scattering 85
Nitrogen 42,43,53
Nuclear
  charge radius 4
  gamma transition 16,17,19,21
  quadrupole moment 4
  spin 3,4,16,17,19,21

Optical
  absorption 19,30,34,48-50
  emission 19
  isotope shift 27
  matrix shift 50,51
Oxygen 61,62,68,83

Paramagnetic relaxation *see also* Spin-
  lattice relaxation 18,20,27
Phonon
  resonant modes 30
  side bands 30
  spectrum 29-31
Phonons 63,64
Phosphorous 43-44
  hydride 43

Photo
  -chemical generation 7
  -dissociation 7,8,21
  -excitation 8
Photolysis 7,8,35,36,42,43
Potassium 34

Quadrupolar interaction 2,5,20-22,
  63-73,79
Quadrupole echo 63,64

Radioactive decay 8
Radiochemical generation 8
Recoil energy 29
Recombination 7-9,24
  time 8,24
Refrigerator 5
Relaxation 60,70,77,86
Relaxation rate (time) 18,31
Rubidium 8,34,39

Satellite transitions 63-65
Self-consistent field (SLF) calculations
  25-27
Self-diffusion 61,77,84-88
Silver 40-42
Sodium 37-39
Solids 59-74,87-92
Spectrometer
  for ESR experiments 5,16
  for NGR experiments 5,10
Spin
  Hamiltonian 2-4,35,37
  effective 3,4,19,27,28
  lattice relaxation 2,4,16,31,33,50
  nuclear 3,4,16,17,19,21,27
  orbit coupling (interaction) 3,52,53

Structure factor 79,80
Substrate 5,10,11,13,15
Super hyperfine interaction 37,40,42,
  45
Surface migration 6,10

T_1 63,66,70-79,87-92
$T_{1\rho}$ 62,67,70-73
T_2 60,61,65-72,78,79,87-91
Tellurium 21,25,27
  dimer 21
  hydride 21
Temperature shift of magnetic hf parameters 47
Tin
  atom 16,17,25-27
  ion 24,25
Tracer measurements 68,74-77
Trapping site see also Lattice site
  13,14,34,37,39,42,45,47-51,53,54
  interstitial 36,40,54
  substitutional 35,36,39,42,47,54

Vacancies 62,68,72
van der Waals
  binding (forces, interaction) 45,
  52,53
  complexes 45

Wave function, overlap 26,52,53

^{129}Xe 60-62,67,76-82
^{131}Xe 60-62,74,78-82

Zeeman splitting
  electronic 3,28,32,49,51
  nuclear 3

Inert Gases

Potentials, Dynamics, and Energy Transfer in Doped Crystals

Editor: **M. L. Klein**
1984. 89 figures. Approx. 266 pages
(Springer Series in Chemical Physics,
Volume 34)
ISBN 3-540-13128-0

Contents: *M. L. Klein:* Argon and Its Companions. - *R. A. Aziz:* Interatomic Potentials for Rare-Gases: Pure and Mixed Interactions. - *S. S. Cohen, M. L. Klein:* Dynamics of Impure Rare-Gas Crystals. - *H. Dupost:* Spectroscopy of Vibrational and Rotational Levels of Diatomic Molecules in Rare-Gas Crystals. - Subject Index.

Y. N. Molin, K. M. Salikhov, K. I. Zamaraev

Spin Exchange

Principles and Applications in Chemistry and Biology

1980. 68 figures, 41 tables. XI, 242 pages
(Springer Series in Chemical Physics,
Volume 8)
ISBN 3-540-10095-4

Contents: Introduction. - Theory of Spin Exchange. - Experimental Measurement of Spin Exchange Rate. - Spin Exchange in Chemistry and Biology. - References. - Subject Index.

Mössbauer Spectroscopy I

Editor: **U. Gonser**
1975. 96 figures. XVIII, 241 pages
(Topics in Applied Physics, Volume 5)
ISBN 3-540-07120-2

Contents: *U. Gonser:* From a Strange Effect to Mössbauer Spectroscopy. - *P. Gütlich:* Mössbauer Spectroscopy in Chemistry. - *R. W. Grant:* Mössbauer Spectroscopy in Magnetism: Characterization of Magnetically-Ordered Compounds. - *C. E. Johnson:* Mössbauer Spectroscopy in Biology. - *S. S. Hafner:* Mössbauer Spectroscopy in Lunar Geology and Mineralogy. - *F. E. Fujita:* Mössbauer Spectroscopy in Physical Metallurgy.

Mössbauer Spectroscopy II

The Exotic Side of the Method

Editor: **U. Gonser**
1981. 67 figures. XII, 196 pages
(Topics in Current Physics, Volume 25)
ISBN 3-540-10519-0

Contents: *U. Gonser:* Introduction. - *R. L. Mössbauer, F. Parak, W. Hoppe:* A Solution of the Phase Problem in the Structure Determination of Biological Macromolecules. - *R. V. Pound:* The Gravitational Red-Shift. - *V. I. Goldanskii, R. N. Kuzmin, V. A. Namiot:* Trends in the Development of the Gamma Laser. - *R. L. Cohen:* Nuclear Resonance Experiments Using Synchrotron Sources. - *U. Gonser, H. Fischer:* Resonance γ-Ray Polarimetry. - *B. D. Sawicka, J. A. Sawicki:* Iron-Ion Implantation Studied by Conversion-Electron Mössbauer Spectroscopy. - *R. S. Preston, U. Gonser:* Selected "Exotic" Applications. - *S. S. Hanna:* The Discovery of the Magnetic Hyperfine Interaction in the Mössbauer Effect of ^{57}Fe.

Springer-Verlag Berlin Heidelberg New York Tokyo